Principles of
Control Engineering

]05

Principles of Control Engineering

Fred White

Senior Lecturer in Electrical Engineering,
Havering College of Further and Higher Education, Essex

Edward Arnold
A member of the Hodder Headline Group
LONDON   SYDNEY   AUCKLAND

First published in Great Britain 1995 by
Edward Arnold, a division of Hodder Headline PLC,
338 Euston Road, London NW1 3BH

Distributed in the USA by
Routledge, Chapman and Hall, Inc.
29 West 35th Street, New York, NY 10001

British Library Cataloguing in Publication Data
A catalogue record for this book is available from the British Library

ISBN 0 340 62541 4

1 2 3 4 5 95 96 97 98 99

Produced by Gray Publishing, Tunbridge Wells
Printed and bound in Great Britain by J. W. Arrowsmith Ltd, Bristol

Contents

Preface

This book has been prepared as a support text for BTEC HNC/HND and undergraduate Linear Control Engineering modules. This book will also be of use to those practising engineers who need to refresh their knowledge or to open-learning students.

Students are often apprehensive of this topic, especially when confronted by differential or integral calculus beyond their experience. I have consciously minimised the use of mathematics to that level experienced by ONC/OND students until well into the text.

I am convinced that all engineering students should have exposure to this subject and an insight into Control Engineering methods will help an overall perception to develop. Another great asset to the engineering student is **CO**ntrol **D**esign **A**nd **S**imulation (CODAS) software, available for the PC, which can be used to demonstrate the methods as they are introduced. Throughout this book I have used CODAS for this purpose and would strongly advise any engineering student to attempt to use it, or at least something similar.

To the mathematician, or even the great Red-Pen himself, I make no apology for leaving out proofs which are not required in most modules and only serve to confuse many students.

Finally, I would like to thank my wife Brenda and my colleagues from Havering College of Further and Higher Education for their support and assistance throughout the writing of this book.

FRED WHITE

For information on CODAS, contact:

Golten and Verwer Partners
33 Moseley Road
Cheadle Hume
Cheshire SK8 5HJ
U.K.

Tel. (0161) 485 5435

1 Introduction

Control Engineering is a fascinating subject which often meets with dismay from students studying it for the first time. The objective of this book is to introduce the student to Control Engineering concepts with a minimum of damage due to mathematical rigour.

The book starts with a brief discussion of the Transfer Function, in which θ is used as a general symbol for a signal with a subscript to show whether input or output. This is itself a function of the s-plane and so should follow an introduction to the Laplace transform. Instead, the reader will be asked to use the s-operator as equivalent to $j\omega$ (this is true as long as the initial conditions are zero) and then solve problems with college-level mathematics.

Some thoughts on the s-plane

One of the major problems faced by engineering and science students is the difficulty of the mathematical methods used to form models and then solve the equations. This is mainly due to the differential calculus involved. Forming differential equations and then solving them for a particular set of values is difficult enough, but we also wish to manipulate these equations in-between forming and solving them. The most common solution to this problem for engineers is to use operator methods. There are several of these methods, but for reasons outlined later in the text we shall be using the Laplace transform or s-operator. This is why the equations used for Transfer Functions are presented in terms of s. I have said in the text that I do not want to start off with pure mathematics, and so the reader is asked to accept equations in this form and only do what is asked, for example put $s = j\omega$, without a detailed a mathematical explanation. I would like the reader to accept the s-operator in the same way as we would accept a logarithm. If there is a difficult task to do, we will often use logs to simplify this task. In this context we will be taking the problem, putting the functions into logs, manipulating the logs, and then putting the logs back into an expression which gives the desired answer. The s-operator will be used in the same way, as we will be:

- taking the problem (as a function of time and hence in the time domain);
- converting into Laplace transforms (now a function of s and in the s-domain or s-plane;
- manipulating as needed, converting back to the time domain for the desired solution.

All of this process should be much clearer once you have finished the chapter on Control Engineering Mathematics, but it would help if you could accept this explanation until then (see Figure 1.1).

Using this equivalence it will be possible to review Transfer Functions, Block Diagram Analysis and Frequency Response (see Figure 1.2).

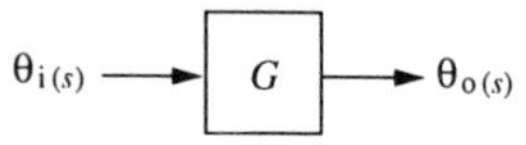

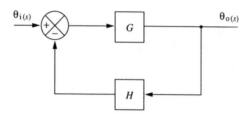

θi(s)
G
θo(s)

Transfer function $G = \frac{\theta_o}{\theta_i}(s)$

Figure 1.1

θi(s)
+
−
G
θo(s)
H

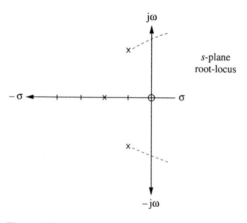

Block diagram with negative feedback loop

Figure 1.2

jω
s-plane
root-locus
−σ
σ
−jω

Figure 1.3

1
0
t
θi
Second order system
θo
1
0
t

Transient response to a step-input

Figure 1.4

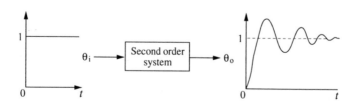

The next topic covered is that of Control Engineering mathematics; this topic often causes problems to students. This is normally because the student does not really have any concept of differential and integral calculus – even though they have passed exams on them. In any case the 'first-principles' content will be kept to a minmum and we will concentrate on problem solving, where the most difficult part will be that on partial fractions.

To complete the s-plane picture let us review Root-locus methods, even though they may not in fact form a part of your control module (see Figure 1.3).

Penultimately, the Bode Standard Second-order equation will be used to obtain information about transient behaviour in the Time Domain (see Figure 1.4).

Ultimately there is a section, Chapter 11, where mixed examples are given followed by worked solutions. In addition to this there are some practical tasks. All of these tasks are simple to perform given the practical resources found in any engineering department.

Any BTEC student who follows the text, works through all the examples and completes the examples in Chapter 11 should be successful in the module.

2

The Transfer Function

The Transfer Function is the output/input signal relationship of a component or system expressed as a function of s. This means that the expressions used to describe the signals must be in terms of the Laplace transform (see Chapter 8 on Control Engineering Mathematics).

As mentioned before, new mathematical techniques will not be introduced until much later in the book. Therefore, we shall signify that the functions used are in terms of s and use $s = j\omega$ where necessary. It is not necessary to derive Laplace transforms or anything of that sort. We will now look at some examples of transfer functions and see how they may be obtained.

❑ **Example 2.1** (see Figure 2.1)

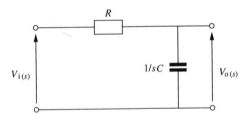

Figure 2.1

The ratio $V_o/V_i(s)$ can be obtained using the voltage divider rule. If you cannot remember this rule then turn to the appendix at the end of this chapter where it is explained

$$\frac{V_o}{V_i}(s) = \frac{\frac{1}{sC}}{R + \frac{1}{sC}}.$$

Multiply all terms by sC

$$\frac{V_o}{V_i}(s) = \frac{1}{1 + sCR}.$$

Remember that CR is the circuit time constant τ, hence this form is a first-order Transfer Function and is often described as

$$\frac{V_o}{V_i}(s) = \frac{1}{1 + s\tau}.$$

We have obtained an expression which describes the network from which it was obtained. Later on you will see that once this function has been obtained we can determine the system response (output signal) to any given type of input signal.

Very often θ (Greek theta) is used to describe the signal, accompanied by the appropriate subscript as to whether the signal is input or output.

❑ *Example 2.2* (see Figure 2.2)

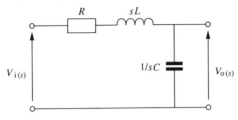

Figure 2.2

To determine this Transfer Function it is necessary to use the voltage divider rule again

$$\frac{V_o}{V_i}(s) = \frac{\dfrac{1}{sC}}{R + sL + \dfrac{1}{sC}}.$$

Multiply all terms by sC

$$\frac{V_o}{V_i}(s) = \frac{1}{s^2 LC + sCR + 1}.$$

This is the Transfer Function, but there is a particular reason (explained in Chapter 9 on Time Response) for making the coefficient of $s^2 = 1$, hence we need to divide all terms by LC. This standard form will be explored several times in the book and so we can leave this in the form here for now.

Apply the principles used in the previous examples to find the Transfer Function of the following networks. Check your answers against the solutions given in the appendix to this chapter.

❑ *Example 2.3* (see Figure 2.3)

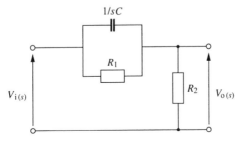

Figure 2.3

❑ *Example 2.4* (see Figure 2.4)

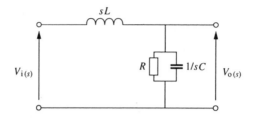

Figure 2.4

Transfer Function of a d.c. generator

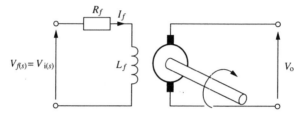

Figure 2.5

The Transfer Function of the d.c. generator shown in Figure 2.5 will relate to the generated voltage (V_o) (open-circuit output voltage) to the field voltage (V_i).

A greatly simplified picture is being given here because the saturation and hysteresis are going to be neglected. Also, all constants will be lumped together into one constant k_g. In reality this constant will convey information about the generator's performance. The unit of the constant k_g will be volts per field ampere.

Kirchhoff's voltage law is applied to the field circuit

$$V_i = L_f \frac{di_f}{dt} + R_f i_f.$$

We can put s equal to the 'differential' using a process described later in Chapter 8 on Control Mathematics

$$V_{i(s)} = sL_f i_{f(s)} + R_f i_{f(s)}$$
$$V_{i(s)} = (sL_f R_f) + i_{f(s)}$$
$$V_{o(s)} = k_g i_{f(s)}.$$

These expressions can be rearranged for $i_{f(s)}$ and then substituted into the equation for $V_{i(s)}$ giving

$$V_{i(s)} = (sL_f + R_f)\frac{V_{o(s)}}{k_{g(s)}}.$$

Hence

$$\frac{V_{o(s)}}{V_{i(s)}} = \frac{\dfrac{k_g}{R_f}}{1 + s\dfrac{L_f}{R_f}},$$

which is of the form

$$\frac{\theta_o}{\theta_i}(s) = \frac{1}{1 + s\tau},$$

where time constant $\tau = L_f/R_f$.

The purpose of this being included at this point is to show that the Transfer Function of any system or component can be obtained if the first principles that apply are known. This one has been used as an example as it is one of the simplest; most electrical engineering students should have a concept of the operation of a d.c. generator.

The next chapter will deal with block diagram analysis. Here the Transfer Functions of feedforward and feedback elements will be represented by G and H, respectively.

Appendix: the Transfer Function

The voltage divider rule

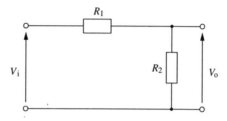

Figure 2.6

In the circuit shown in Figure 2.6 the current (I) is

$$\frac{V_i}{R_1 + R_2}$$

and the output voltage (V_o) is this current multiplied by the resistor R_2

$$V_o = \frac{R_2 V_i}{R_1 + R_2}$$

and hence the ratio

$$\frac{V_o}{V_i} = \frac{R_2}{R_1 + R_2}.$$

This is an important rule since it can be used in almost all circuits.

Solution to Example 2.1

The circuit components are made functions of s which, at the moment, simply means putting $s = j\omega$.
   Remember that

$$X_c = \frac{1}{2\pi f C}, \quad = \frac{1}{j\omega C} = \frac{1}{sC}.$$

Hence, the circuit impedance

$$Z = R + \frac{1}{sC}$$

giving circuit current

$$I = \frac{V_i}{R_1 + \dfrac{1}{sC}}$$

with output voltage

$$V_o = \frac{\dfrac{V_i}{sC}}{R + \dfrac{1}{sC}}$$

and hence, the Transfer Function

$$\frac{V_o}{V_i}(s) = \frac{\dfrac{1}{sC}}{R + \dfrac{1}{sC}}.$$

To tidy this expression up multiply all terms by sC to give

$$\frac{V_o}{V_i}(s) = \frac{1}{1 + sCR}.$$

This is the standard form for the Transfer Function of a first-order system (see Chapter 9 on Time Response), where CR is the time constant of that system.

Solution to Example 2.3

First find the equivalent impedance to R_1 and $1/sC$ in parallel. This is simply achieved using the product over sum technique

$$\frac{R_1 \cdot \dfrac{1}{sC}}{R_1 + \dfrac{1}{sC}} = \frac{\dfrac{R_1}{sC}}{R_1 + \dfrac{1}{sC}} = \frac{R_1}{1 + sCR_1}.$$

This can now be used as part of the series impedance and then used to apply the voltage divider rule, i.e.

$$\frac{V_o}{V_i}(s) = \frac{R_2}{R_2 + \dfrac{R_1}{1 + sCR_1}}.$$

Multiply all terms by $(1 + sCR_1)$ and

$$\frac{V_o}{V_i}(s) = \frac{R_2(1+sCR_1)}{R_1+R_2+sCR_1R_2}.$$

Solution to Example 2.4

The parallel impedance has an equivalent value as in the previous example

$$\frac{R}{1+sCR}.$$

This is used to determine the Transfer Function using the voltage divider rule

$$\frac{V_o}{V_i}(s) = \frac{\dfrac{R}{1+sCR}}{sL+\dfrac{R}{1+sCR}}$$

multiply all terms by $(1 + sCR)$

$$\frac{V_o}{V_i}(s) = \frac{R}{sL(1+sCR)+R}$$

$$= \frac{R}{s^2LCR+sL+R}.$$

Later in the book we will want to take this into another form, with the coefficient of $s^2 = 1$, however, at the moment it is not necessary to explain it further.

3 Block Diagram Analysis

We want to model real systems and so be able to select individual components of the system and then identify how they will interact. We have seen that the Transfer Function of the component provides information which will identify its characteristic, and we now want to analyse a complete system, or sub-system from its parts.

In this process, each element of the control system is treated as a black box with a Transfer Function (TF) which is defined as the ratio of

$$\frac{\theta_o}{\theta_i},$$

where θ represents a signal, electrical or otherwise. The subscripts 'o' and 'i' signify output and input, respectively.

An input signal is also referred to as a **cause** and an output signal as an **effect**. Therefore the output–input relation is also known as the cause and effect relation.

This may be better illustrated by some examples of transfer functions or **cause and effect** relationships.

❑ *Example 3.1*

A voltage V applied to a resistor R causes a current I to exist. Remembering that current is the rate of flow of charged particles. The output–input relation is $I = V/R$. In the circuit the input is V and the output is I.

The black boxes used in block diagram representations are shown with the box representing the element and arrows indicating the flow of information through the system. Figure 3.1 shows the block diagram representation of Ohm's law.

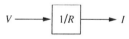

Figure 3.1

❑ *Example 3.2* (see Figure 3.2)

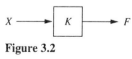

Figure 3.2

A spring has a resisting tensile force of F which is proportional to its extension X, so that $F = kX$ (where k is the spring constant)

$$k = \frac{F}{x}.$$

❏ *Example 3.3* (see Figure 3.3)

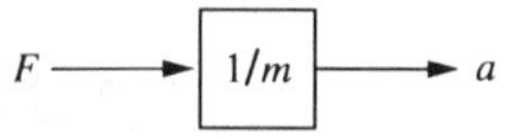

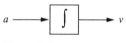

Figure 3.3

Newton's law states that a force applied to a mass m causes an acceleration a, hence $F = ma$.

Black box representation can be used to represent a process as well as a component or element. For instance integrating acceleration a over time produces the velocity v, i.e.

$$v = \int a \cdot dt.$$

Therefore acceleration is the cause of velocity. This process can be represented in black box form as shown in Figure 3.4.

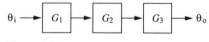

Figure 3.4

Elements connected in cascade

When dealing with general form block diagrams, we represent the forward gain of the element by G. If there is more than one element then they can be numbered G_1, G_2, etc. To determine the overall forward gain we multiply the individual gains (or Transfer Functions)

$$G_1 \times G_2 \times \text{etc.}$$

❏ *Example 3.4*

An open-loop control system can be represented as shown in Figure 3.5; to determine the overall system Transfer Function.

θ$_i$ → G_1 → G_2 → G_3 → θ$_o$

Figure 3.5

We have already stated that the **feedforward Transfer Function**, or gain, is a product of the Transfer Functions of each element in the feedforward path or

$$G = G_1 G_2 G_3$$

and from the cause and effect principle we can deduce that

$$\theta_o = \theta_i G_1 G_2 G_3$$

hence the open-loop Transfer Function

$$\frac{\theta_o}{\theta_i} = G_1 G_2 G_3.$$

Closed-loop systems: feedback and stability

Hopefully, you will already be familiar with the concept of feedback in electronics. If you are not you will soon cover this topic in your course.

Let us consider the (hypothetical) example of someone driving a motorcar on an empty motorway (should one exist). If the driver were to keep her foot at a constant angle on the accelerator, then the car would have a constant input and only change speed as the incline of the road altered.

Normally, of course, we drive with our eyes open and senses operating such that road and traffic information is fed into the brain via our eyes and ears and the necessary adjustments are made to the engine speed. In this case the feedback loop is by way of the brain and driving experience.

The open-loop systems would have a fixed input setting and the output would be a function of variations in the load of whatever process we are controlling. In other words, an increase in load would mean that the system slowed down and vice versa.

The closed-loop system would contain an element, or a number of elements, which will measure the output (speed, position or whatever) and compare this to a reference and then make the necessary adjustment to compensate for deviation, or error, from the desired output.

The feedback elements will have transfer functions which (in Control Engineering) are given the symbol H. Feedback may be fractional or unity (100%) depending on the type of system.

A unity feedback system is one where the nature of the feedback signal is of the same nature as the input signal. For example, if a system has an electrical input and the feedback is by way of the output of a tacho-generator, then both input and feedback signals are electrical and can be compared at the summing device without the need for an additional transducer in the feedback path. The feedback system will require a summing junction (comparator) which will allow the feedback signal to be added or subtracted from the input, depending on whether the feedback is positive or negative. The polarity of the comparator is indicated in the symbol.

Positive feedback with feedback Transfer Function H

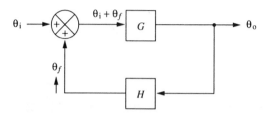

Figure 3.6

In Figure 3.6 θ_i is the input signal, θ_o is the output signal and H is the feedback transfer function, $\theta_f = \theta_o H$, therefore

$$\text{without feedback } \theta_o = \theta_i G \qquad \text{with feedback } \theta_f = \theta_o H$$

$$\text{and } (\theta_i + \theta_f)G = \theta_o \qquad \text{therefore } (\theta_i + \theta_o H)G = \theta_o$$

$$G\theta_i + G\theta_o H = \theta_o$$

$$G\theta_i = \theta_o - G\theta_o H$$

$$G\theta_i = \theta_o(1 - GH)$$

and

$$\frac{\theta_o}{\theta_i} = \frac{G}{1 - GH}.$$

A similar exercise can exact the Transfer Function for negative feedback or unity feedback of either polarity, but at this point just remember the standard form

$$\frac{\theta_o}{\theta_i} = \frac{G}{1 \pm GH}.$$

An important condition occurs with positive feedback when GH is equal to one as this brings the denominator to zero and hence the overall value of the Transfer Function to infinity. Obviously this would be highly undesirable in any Control System. This will be considered in more detail when we consider damping and system time response in Chapter 9.

❑ *Example 3.5*

A Control System has a forward gain of 20 and the proportion of the output which is feedback is 0.04 (4%). Calculate the closed-loop Transfer Function when (a) positive feedback is used and (b) negative feedback is used.

❏ *Example 3.6*

For the system used in Example 3.5 calculate the closed-loop gain if the forward gain is reduced by (a) 25% and (b) 50%.

❏ *Example 3.7*

With the forward gain set at 20 (first positive feedback then negative) and with the feedback fraction set at 5% of θ_i what will the closed-loop gain be now?

Summary

We can now summarise the salient points.

Positive feedback

Here the advantage is that the introduction of a feedback loop will result in an increase in gain but with the danger that the output could become unstable. If oscillations are required at the output then this may be desirable, but generally it is not.

Negative feedback

A stable output is assured after a stable system has been designed, however, further stages in the feedforward path may be required to achieve the required closed-loop gain.

We will now look at various methods for combining multi-block or multi-loop systems.

Moving the take-off point

❑ *Example 3.8* (see Figure 3.7)

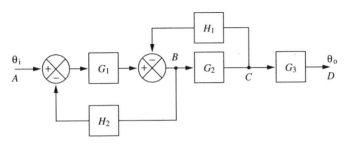

Figure 3.7

Elements in series can only be combined when there are no summing units or loop take-off points between them. Similarly, feedback loop reduction in parallel systems can only occur when there are no other take-off points or feedback loops within the loop.

Consider the above example: we need to move the take-off point for the lower loop from B to C thus freeing the upper H_1 loop. This can be achieved as shown below.

We need to add an element $1/G_2$ to the lower loop to change B input to C, this additional element will not affect the analysis and so ensure that we get the desired result.

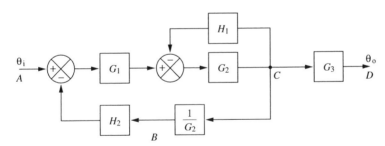

Figure 3.8

The system shown in Figure 3.8 can now be analysed in the normal way to give

$$\frac{\theta_o}{\theta_i} = \frac{G_1 G_2 G_3}{1 + G_2 H_1 + G_1 H_2}.$$

Summary of Block Diagram transformations

Figure 3.9(a)–(i) presents a summary of Block Diagram transformations.

(a)

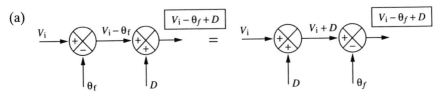

(b)

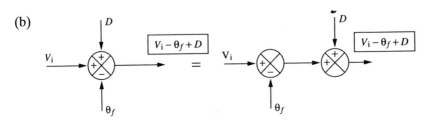

(c)

θ_i → G_1 → G_2 → θ_o = θ_i → $G_1 G_2$ → θ_o

(d)

θ_i → G_1 → ⊗ → θ_o
         G_2
= θ_i → $G_1 + G_2$ → θ_o

(e)

θ_i → G → ⊗ → θ_o
              D
= θ_i → ⊗ → G → θ_o
              $1/G$ ← D

(f)

θ_i → ⊗ → G →
         θ_f
= θ_i → G → ⊗ →
   θ_f → G

Figure 3.9(a)–(f)

(g)

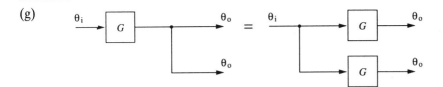

(h)

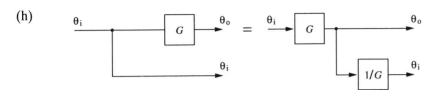

(i)

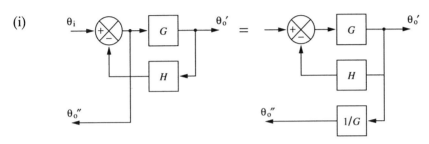

Figure 3.9(g)–(i)

Multiple input, multiple output (MIMO) systems

In practical systems we will frequently encounter multiple inputs, often due to disturbance of the measurement of more than one parameter. To analyse these systems we need to use **superposition theorem**, a wonderfully complicated-sounding title for what is a simple technique.

❏ *Example 3.9*

In the system shown in Figure 3.10 we want to find an expression for the output signal in terms of input θ_i, and disturbance D. We cannot do this in a single pass, so we take each input in turn, making the other zero. Each input is then transposed to give an expression for θ_o, we then superimpose or add the outcome of each. Note that unity feedback has been used which means $H = 1$.

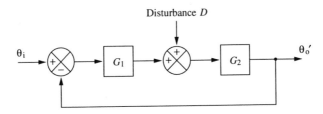

Figure 3.10

Firstly, we can consider $D = 0$ and so the transfer function is

$$\frac{\theta_o}{\theta_i} = \frac{G_1 G_2}{1 + G_1 G_2}.$$

This is only part of θ_o and we will call it θ_o'.

$$\theta_o' = \frac{\theta_i G_1 G_2}{1 + G_1 G_2}.$$

Now, if we consider the disturbance D as the input and $\theta_i = 0$, we can represent the system in this form shown in Figure 3.11.

Note that the G_1 feedback block carries through the negative sign and changes the sign in the remaining summing junction. As the original input θ_i has a value of zero then the transfer function θ_o''/D is derived

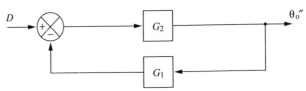

Figure 3.11

$$\frac{\theta_0''}{D} = \frac{G_2}{1+G_1G_2} \quad \text{or} \quad \theta_0'' = \frac{DG_2}{1+G_1G_2}.$$

When both θ, and D have measurable values, we then need to combine or **superimpose** the two solutions because $\theta_o = \theta_0' + \theta_0''$

$$\theta_o = \theta_0' + \theta_0'' = \frac{\theta_o G_1 G_2}{1+G_1G_2} + \frac{DG_2}{1+G_1G_2}$$

$$= \theta_o = \frac{G_2}{1+G_1G_2}(\theta_i G_1 + D).$$

Block diagram reduction examples

❏ *Example 3.10*

Reduce the block diagram shown in Figure 3.12 and produce a system Transfer Function.

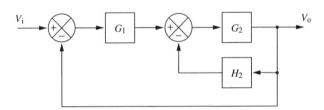

Figure 3.12

❏ *Example 3.11*

Use the superposition theorem to determine an expression for θ_o when both D and θ_i have non-zero values (see Figure 3.13).

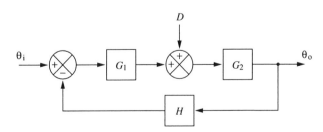

Figure 3.13

❑ *Example 3.12*

Determine the overall Transfer Function of the system shown below. It will be to your advantage in the first instance to move the take-off point of H_1 to the output of G_2 (see Figure 3.14).

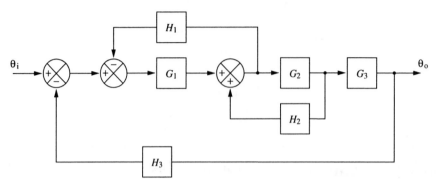

Figure 3.14

Appendix: solutions to Block Diagram problems

❏ *Example 3.5(a)* (see Figure 3.15)

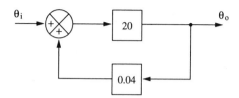

Figure 3.15

Using the relationship derived in the text

$$\frac{\theta_o}{\theta_i} = \frac{G}{1 - GH} = \frac{20}{1 - 0.8} = 100.$$

❏ *Example 3.5(b)* (see Figure 3.16)

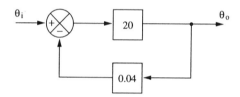

Figure 3.16

$$\frac{\theta_o}{\theta_i} = \frac{G}{1 + GH} = \frac{20}{1 + 0.8} = 11.1^r.$$

❏ *Example 3.6(a)*

Positive feedback: see Figure 3.17

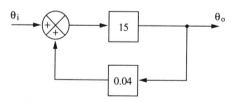

Figure 3.17

$$\frac{\theta_o}{\theta_i} = \frac{15}{1 - (0.04 \times 15)} = 37.5.$$

This is a 62.5% reduction in overall gain for a 25% reduction in component gain.
  Negative feedback: see Figure 3.18

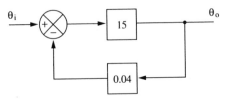

Figure 3.18

$$\frac{\theta_o}{\theta_i} = \frac{15}{1+(0.04\times15)} = 9.375.$$

Here we have a 15.6% reduction in overall gain for the same 25% reduction in component gain.

❏ *Example 3.6(b)*

Positive feedback: see Figure 3.19

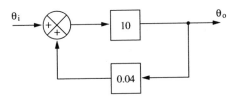

Figure 3.19

$$\frac{\theta_o}{\theta_i} = \frac{10}{1-(0.04\times10)} = 16.67.$$

This is an 83.3% reduction in overall gain for a 50% reduction in component gain.
  Negative feedback: see Figure 3.20

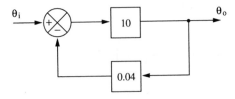

Figure 3.20

$$\frac{\theta_o}{\theta_i} = \frac{10}{1+(0.04\times10)} = 7.1428.$$

This shows only a 35.7% reduction in overall gain for a 50% reduction in component gain. This is an important characteristic to note of negative feedback.

❑ *Example 3.7*

Positive feedback: see Figure 3.21

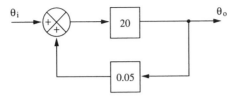

Figure 3.21

$$\frac{\theta_o}{\theta_i} = \frac{20}{1-(0.05 \times 20)} = \text{infinite}$$

this is an unstable output.

Negative feedback: see Figure 3.22

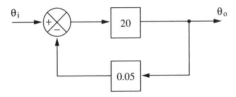

Figure 3.22

$$\frac{\theta_o}{\theta_i} = \frac{20}{1+(0.05 \times 20)} = 10$$

this is a stable output.

❑ *Example 3.8*

Once the take-off point has been moved (see the summary of Block Diagram transpositions, pages 18–19) then we can analyse in the usual manner (see Figures 3.23 and 3.24).

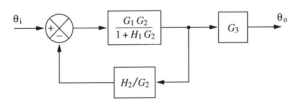

Figure 3.23

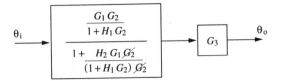

Figure 3.24

$$\frac{\theta_o}{\theta_i} = \frac{G_1 G_2 G_3}{1 + G_2 H_1 + G_1 H_2}.$$

❑ *Example 3.9*

The explanation should be clear from the description in the text.

❑ *Example 3.10* (see Figures 3.25 and 3.26)

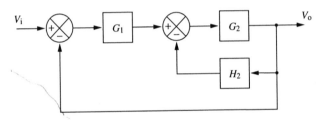

Figure 3.25

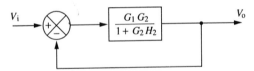

Figure 3.26

$$\frac{V_o}{V_i}(s) = \frac{G_1 G_2}{1 + G_2 H_2 + G_1 G_2}.$$

❑ *Example 3.11*

When θ_i = input and $D = 0$ (see Figure 3.27), then

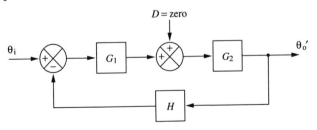

Figure 3.27

$$\frac{\theta_0'}{\theta_i} = \frac{G_1G_2}{1+G_1G_2H} \quad \text{and} \quad C' = \frac{\theta_i G_1 G_2}{1+G_1G_2H}.$$

When D = input and $\theta_i = 0$ (see Figure 3.28), then

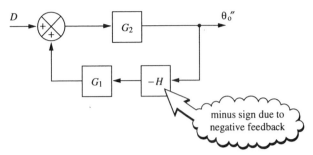

Figure 3.28

$$\frac{\theta_0''}{D} = \frac{G_2}{1+G_1G_2H} \quad \text{and} \quad \theta_0'' = \frac{DG_2}{1+G_1G_2H}$$

$$\theta_0 = \theta_0' + \theta_0'' = \frac{\theta_i G_1 G_2 + DG_2}{1+G_1G_2H}$$

$$= \frac{G_2(\theta_i G_1 + D)}{1+G_1G_2H}.$$

❏ ***Example 3.12*** (see Figures 3.29–3.31)

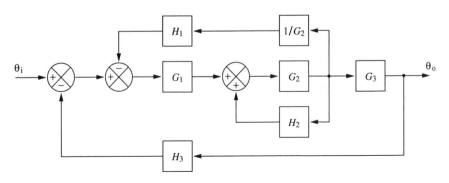

Figure 3.29

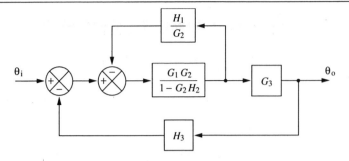

Figure 3.30

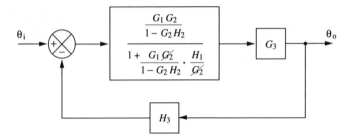

Figure 3.31

The overall Transfer Function is given as

$$= \frac{G_1 G_2 G_3}{1 - G_2 H_2 + G_1 H_1 + G_1 G_2 G_3 H_3}.$$

4 Frequency response
I. The Nyquist diagram

A review of complex numbers

If we consider a complex number in which one (or both) parts are variable, then as they vary so the number may vary in both gain and phase (magnitude and argument). In the frequency response test the results represent a frequency dependent phasor which we can express in terms of $j\omega$ (where ω is angular frequency). When ω varies the gain and phase described above varies.

In a Control system any frequency dependent component will have a complex value expressed by angular frequency omega (ω). Recall that angular frequency is measured in radians per second and

$$\omega = 2\pi f \text{ rad/s.}$$

❏ *Example 4.1*

Take a complex number $1 + j2\omega$ and calculate its value in magnitude and phase for values of $\omega = 2, 4, 6, 8$ and 10 rad/s

$$\omega = 2, \text{ then } 1 + j4 = \sqrt{17} \text{ angle } (\tan^{-1} 4)$$

which equals $\sqrt{17}$ angle (75.96°) continue for $\omega = 4, 6, 8$ and 10 rad/s and check your answers with those in the appendix (page 35).

This is the calculation process required for the Nyquist diagram. For practice complete the examples given below.

❏ *Example 4.2*

For the angular frequencies given in Example 4.1 calculate the magnitude and phase for each of the following complex numbers

(a) $\dfrac{10}{j\omega(1+j\omega)}$

(b) $\dfrac{1+j\omega}{1+j5\omega}$

(c) $(1+j\omega)(1+j2\omega)$

(d) $\dfrac{100}{(1+j\omega)(1+j5\omega)}.$

The solutions to these exercises are given in the appendix at the end of the chapter (page 35).

Frequency response

Previously we have seen that if we know the Transfer Functions of component parts of a system, then the system can be analysed to determine its overall Transfer Function. The reason for doing this may not be clear yet. However, it is easy to appreciate that at the design stage one does not want to have to build a succession of modified systems in order to eventually obtain the required performance. A much more straightforward method is to construct a mathematical model (often using a computer), which can, subjected to various sets of inputs, disturbances and so on, produce data to model the system's performance.

Often the Transfer Function may not be known; this and other criteria can be found from the test data. The system can also be used to determine how the system will operate when the feedback loop is closed, without actually having to close it.

There are several techniques which are used to these ends, such as the frequency response methods introduced earlier in this chapter.

Software is readily available which can carry out all of the calculations contained in this chapter in just a few seconds. User-friendly software now makes it easy to solve these types of problems with the minimum of instruction. However, it is still necessary for the prospective engineer to have a good concept of the processes involved. Hence, we shall go through the methods mentioned using examples to demonstrate the techniques involved.

The Nyquist diagram

The Nyquist diagram is a polar plot of the open-loop frequency response (V_o/V_i). This is obtained by subjecting the plant (load and controller) to a sinusoidal input of fixed magnitude ($1V$ is very useful as V_o is now the Transfer Function) but of varying frequency. The Transfer Function (V_o/V_i) is then plotted on polar graphpaper with angular velocity increasing clockwise. To achieve this the output must be observed both in magnitude and phase; we can easily do this and a practical, task for this purpose is given in the appendix.

We have seen that transfer functions are normally presented in terms of the s-operator and for the purpose of the Nyquist diagram we replace s by $j\omega$ (where ω is $2\pi f$ in radians per second). If we take a simple example of a system with an open-loop transfer function

$$G(s) = \frac{5}{s(1+0.2s)}$$

for s we can put $j\omega$ ($\omega = 2\pi f$ rad/s) and evaluate the gain and phase angle for the function for various values of ω

$$G(j\omega) = \frac{5}{j\omega(1+j0.2\omega)}.$$

The modulus of this function

$$\left|G(j\omega)\right| = \frac{5}{\omega\sqrt{1+(0.2\omega)^2}}.$$

and the phase of $G(j\omega) = -(90 + (\tan^{-1} 0.2\omega))°$. Remember that the $j\omega$ outside of the brackets in the denominator will add a constant $90°$ to the phase, and that phase will change its sign when the taken to the numerator.

ω (rad/s)	$|G|$	Angle G
0.5	9.9	−95.71°
2.0	2.32	−111.9°
4.0	0.98	−128.7°
8.0	0.33	−148°
12.0	0.16	−157.4°
16.0	0.093	−162.7°
20.0	0.06	−166°

We see that as ω increases then the angle becomes more negative and the magnitude reduces toward zero. If these values are plotted on polar graphpaper with ω increasing in a clockwise direction the resulting plot is called Nyquist diagram (see Figure 4.1).

Note that the process for plotting the polar values is exactly the same as that of the Archimedes spiral as found in some level 3 mathematics textbooks.

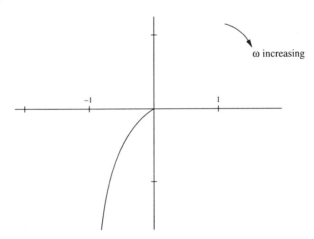

ω increasing

Figure 4.1

❑ *Example 4.3*

Given that a system has an open-loop transfer function

$$\frac{\theta_{o(s)}}{\theta_{i(s)}} = \frac{25}{s(1+0.02s)(1+0.15s)}$$

produce a Nyquist diagram for values of ω between 0.5 and 50 rad/s.

Firstly we need to replace s by $j\omega$ and determine the value of the transfer function both in magnitude and phase.

In order to set out the results in a uniform manner use a table which will allow the calculations to be done in simple steps; use the following 'labels':

■ the phase-angle of the function

$$\frac{\theta_{o(s)}}{\theta_{i(s)}} = -(90 + A + B)^\circ,$$

where the constant 90° is due to the angle of $j\omega$ outside of the brackets,

$$A = \text{angle } (1 + 0.02j\omega) = \tan^{-1} (0.02\omega)^\circ$$

$$B = \text{angle } (1 + 0.15j\omega) = \tan^{-1} (0.15\omega)^\circ.$$

■ The magnitude of the function

$$\left|\frac{\theta_o}{\theta_i}\right| = \frac{25}{\omega CD},$$

where

$$\omega = |j\omega|$$

$$C = |1 + 0.02j\omega| = \sqrt{1 + (0.02\omega)^2}$$

$$D = |1 + 0.15j\omega| = \sqrt{1 + (0.15\omega)^2}.$$

Now set out a table and carry out the calculations, the solution and a sample diagram is in the appendix at the end of this chapter (page 39). Once you have a complete set of values we can plot these on polar graphpaper with omega (ω) increasing in a clockwise direction.

Stability, phase and gain margins

The Nyquist stability criterion tells us that the system will have a stable closed-loop response if the locus crosses the $-180°$ axis ot the right of the -1 point. Whether this occurs or not is a function of the magnitude of the transfer function or the phase of it, or both. Consequently it is important to be able to tell how much 'margin' there is for either gain or phase to increase before a stable system becomes unstable.

What is meant by stable?

We are concerned that an input of energy will cause the system output to oscillate and that these oscillations will continue to grow in magnitude until the system is damaged.

A stable system may also have an oscillating output following a sudden input, but these oscillations will settle down to an acceptable limit. A simple analogy is that a stable system is represented by a ball in a bowl; a schematic of this system is shown in Figure 4.2.

Figure 4.2

If the ball is subjected to an input of energy it will run up the wall of the bowl and then down and up the other side, but the energy will dissipate and the ball will eventually settle back at the bottom of the bowl.

If the system is unstable then we would model it by a ball on an upturned bowl, as shown in Figure 4.3. Any input of energy here would certainly destroy the system.

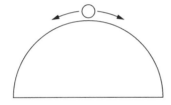

Figure 4.3

Establishing the gain and phase margins

When the Nyquist locus passes through the −180° axis it is possible to measure the distance between the intersection and the −1 point. If this distance x then the gain margin (g_m) is $1/x$ or $-20 \log x$ dB. Also if the point at which the locus crosses a circle of radius 1 is located, then draw a line between that point and the origin, an angle will be made with the −180° axis. This angle is called the phase margin (Φ_m). Both the gain margin and the phase margin will tell us by how much we can increase gain or phase and remain stable.

Let us refer to the Nyquist diagram of Example 4.3 and determine both gain and phase margins (see Figure 4.4). The system gain margin is $1/0.443 = 2.257$ and the phase margin is $180° - 164.2° = 15.8°$.

❏ *Example 4.4*

As an exercise calculate the appropriate phasor values for the transfer functions below, and determine gain and phase margins where possible. In each case use values of angular velocity between 0.5 and 20 rad/s

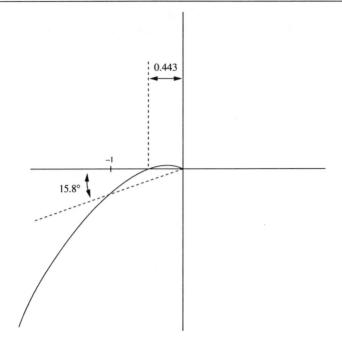

Figure 4.4

(a) $\dfrac{2}{1+5s}$

(b) $\dfrac{2}{s(1+5s)}$

(c) $\dfrac{20}{s(1+s)(2+2s)}$

(d) $\dfrac{30}{(s+0.1)(s^2+6s+8)}$.

Appendix: solution of Nyquist diagram problems

❏ Example 4.1

ω	2	4	6	8	10
$\tan^{-1}2\omega$	75.96°	82.9°	85.24°	86.4°	87.14°
$\sqrt{(1 + [2\omega]^2)}$	$\sqrt{17}$	$\sqrt{65}$	$\sqrt{145}$	$\sqrt{257}$	$\sqrt{401}$

❏ Example 4.2

(a)

ω	2	4	6	8	10
$-(90 + \tan^{-1}\omega)$	$-153°$	$-166°$	$-171°$	$-173°$	$-174.3°$
$\left\lvert\dfrac{10}{\omega(1+\omega^2)}\right\rvert$	$\sqrt{5}$	0.606	0.274	0.16	0.1

(b)

ω	2	4	6	8	10
$\tan^{-1}\omega - \tan^{-1}5\omega$	$-20.9°$	$-11.2°$	$-7.6°$	$-5.7°$	$-4.6°$
$\dfrac{\sqrt{1+\omega^2}}{\sqrt{1+[5\omega]^2}}$	0.23	0.21	0.203	0.202	0.201

(c)

ω	2	4	6	8	10
$\tan^{-1}\omega + \tan^{-1}2\omega$	139°	159°	166°	170°	171.4°
$\{\sqrt{(1 + \omega^2)}\} \times \{\sqrt{(1 + 4\omega^2)}\}$	9.22	33.24	73.3	129.3	201.3

(d)

ω	2	4	6	8	10
$-(\tan^{-1}\omega + \tan^{-1}5\omega)$		$-147.7°$ $-163°$	$-169°$	$-171.4°$	$-173°$
$\dfrac{100}{\{\sqrt{(1+\omega^2)}\sqrt{(1+25\omega^2)}\}}$	4.5	1.2	0.56	0.31	0.2

❑ *Example 4.3*

$$\frac{25}{j\omega(1+j0.02\omega)(1+j0.15\omega)}$$

ω	Angle	Modulus
0.5	−94.9°	49.96
1	−99.7°	24.72
1.5	−104.4°	16.25
2	−109°	11.96
3	−117.7°	7.59
5	−132.6°	3.98
10	−157.6°	1.36
20	−183.4°	0.37
30	−198.4°	0.16
40	−209.2°	0.08
50	−217.4°	0.05

Also see Figure 4.4.

❑ *Example 4.4*

(a) $g_m = \infty$, $\Phi_m = 120°$ (see Figure 4.5).

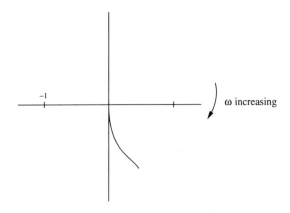

Figure 4.5

(b) $g_m = \infty$, $\Phi_m = 58°$ (see Figure 4.6).

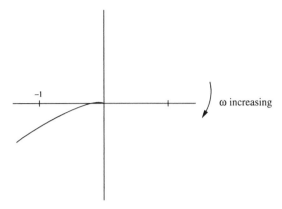

Figure 4.6

(c) Unstable (see Figure 4.7).

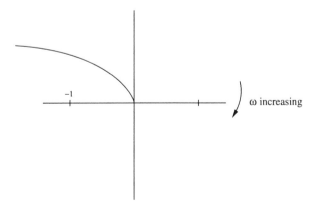

Figure 4.7

(d) $g_m = 1.725$, $\Phi_m = 58°$ (see Figure 4.8).

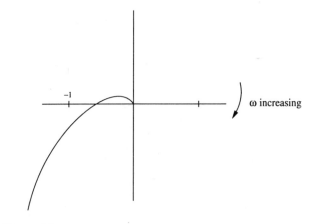

Figure 4.8

5 Frequency response
II. The Bode plot

We have seen how to produce Nyquist diagrams and after only a little practice we can see that unless a computer is used for the calculations the process is protracted. A more rapid technique is to use the Bode plot.

The Bode plot is a representation of open-loop frequency response using logarithmic/linear scales. This method can be very fast when 'straight line approximation' (using asymptotes) is employed.

We will look first at the formal method, this requires some calculation and plotting of points and, though easily achievable by hand, is better suited to computer methods. In this method the modulus of the open-loop transfer function, expressed in decibels, i.e. $20 \log_{10} |\theta_o/\theta_i|$, and the phase of the open-loop transfer function, i.e. angle θ_o/θ_i, are plotted as separate points on a linear scale against log angular frequency (ω).

The angular frequency axis (y-axis) is normally logarithmic to base ten along which equal intervals correspond to an angular frequency ration of 10:1, or a decade. Let us use Example 4.3, where the calculations have already been carried out, and simply plot these (see Figure 5.1).

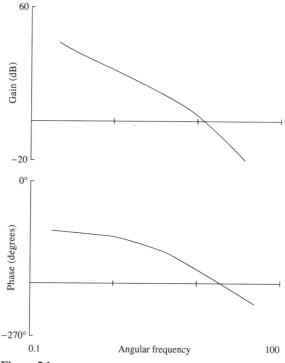

Figure 5.1

Advantages of Bode diagrams

1. Graphs associated with commonly encountered Transfer Functions are easily determined, especially when straight line approximations are used.
2. Multiplication and division of Transfer Functions is simplified due to the use of logarithms (remembering that the product of two numbers is the addition of their logarithms and their quotient is the difference of their logarithms).
3. The phase is related to the slope of the log-modulus characteristic (dB per decade).
4. Design of series compensators is made more rapid (we will not be examining the Nichol's chart and so this may not be obvious in this book).
5. The maximum error when using a straight line approximation is ±3 dB, which always occurs at the break frequency (corner frequency).

Graphs of common Transfer Functions

In Control Engineering most Transfer Functions are a composite of commonly encountered terms, either in the numerator or the denominator. The trick is to be able to recognise these terms and sketch their straight line approximations or asymptotes. The most common terms are

$$(j\omega), \quad (1 + j\omega\tau), \quad (k_1(j\omega)^2 + k_2(j\omega) + 1)$$

and these may in turn be raised to integral powers.

$GH(j\omega) = j\omega$

Consider

$$20 \log_{10}|H(j\omega)| = 20 \log_{10}\omega.$$

The log-modulus (in decibels) for various values of ω is given as

ω (rad/s)	$20 \log 10 |GH(j\omega)|$ (dB)
0.01	−40
0.1	−20
1.0	0
10.0	+20
100.0	+40

The phase angle of $GH(j\omega) = 90°$ for all frequencies and hence the Bode diagram for this function is given as shown in Figure 5.2. The log-modulus graph for this

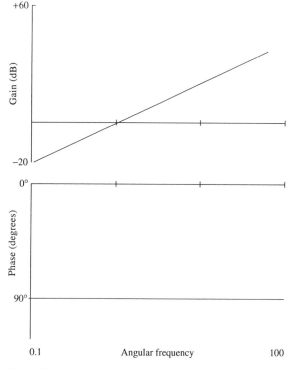

Figure 5.2

function is thus a straight line of slope $+20$ dB/decade. You should be able to produce the Bode diagram for the function $1/(j\omega)$ which gives a straight line of slope -20 dB/decade and a constant phase at $-90°$.

To help sketch this characteristic it is useful to mark the point at which the line crosses 0 dB. Remember that at 0 dB the function has a magnitude of 1 ($\log_{10} 1 = 0$). For $1/j\omega$ to have a magnitude of 1 when $\omega = 1$. If the function was $120/\omega$ then $\omega = 120$ at the cross-over frequency (see Figure 5.3 overleaf).

In general, the function $H(j\omega)^n$ has a log-modulus graph which is a straight line of slope $20n$ dB/decade, where n is a positive or negative integer.

$H(j\omega) = (1 + j\omega\tau)$

Consider

$$20 \log_{10} |H(j\omega)| = 20 \log_{10}\sqrt{(1 + \omega^2\tau^2)}$$

for

$$\omega\tau \ll 1, \sqrt{(1 + \omega^2\tau^2)}$$

is approximately equal to 1, thus

$$20 \log_{10} |H(j\omega)| = 0 \text{ dB}$$

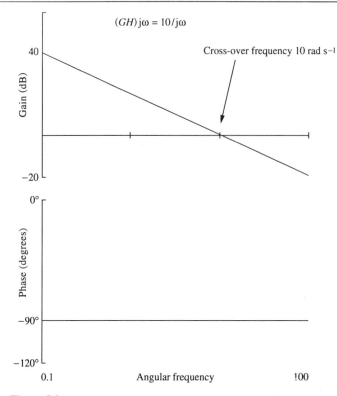

Figure 5.3

for

$$\omega\tau = 1, \sqrt{(1 + \omega^2\tau^2)} = \sqrt{2}$$

thus

$$20 \log_{10}|H(j\omega)| = +3 \text{ dB}$$

for

$$\omega\tau >> 1, \sqrt{(1 + \omega^2\tau^2)}$$

approximately equal to $\omega\tau$, thus

$$20 \log_{10}|H(j\omega)| = 20 \log_{10}\omega\tau \text{ dB}.$$

Hence for low frequencies the log-modulus characteristic is approximately a straight line lying along the 0 dB axis. For high frequencies, the log-modulus is approximately a straight line of slope 20 dB/decade, since for a decade change in frequency the log-modulus changes by 20 dB (see Figure 5.4).

The phase angle of $|H(j\omega)| = \tan^{-1}\omega\tau$ and increases from zero at zero frequency to 90° at infinite frequency. The log-modulus curve is asymptotic to the frequency axis

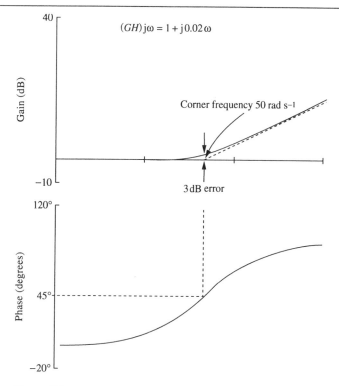

$(GH)j\omega = 1 + j0.02\omega$

Corner frequency 50 rad s^{-1}

3 dB error

Figure 5.4

for low frequencies and is asymptotic for high frequencies to the straight line starting at $\omega = 1/\tau$ with a slope of 20 dB/decade. Furthermore, the error in using two asymptotes instead of the actual curve is never more than 3 dB. This error occurs at $\omega = 1/\tau$ and is termed the **break frequency** or **break point** or **corner frequency** (ω_c).

For most practical purposes asymptotic approximation is sufficient and saves calculation time.

$H(j\omega) = 1/(1 + j\omega\tau)$

The function $H(j\omega) = 1/(1 + j\omega\tau)$ will also have a log-modulus characteristic asymptotic to the 0 dB axis for low frequencies and the break point occurs at $1/\tau$. At high frequencies, the gain is asymptotic to a line of −20 dB/decade. The phase of angle $|H(j\omega)|$ varies from zero to −90° as the frequency changes from zero to infinity.

For functions of the form $H(j\omega) = (1 + j\omega\tau)^n$, where n is a positive integer, the log-modulus is again asymptotic to the 0 dB axis at very low frequencies. The break again occurs at $\omega = 1/\tau$ and the high frequency asymptote has a slope of $+20n$ dB/decade. The phase varies from zero to $90n°$ as the frequency changes from zero to infinity. When n is a negative integer then the slope of the log-modulus asymptote is $-20n$ dB/decade beyond the break point and the phase tends to −90° at infinite frequency (see Figure 5.5).

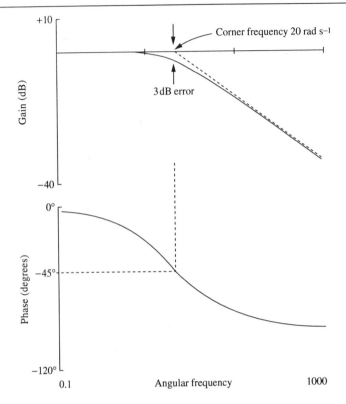

Figure 5.5

$$H(j\omega) = 1/\{k_1(j\omega)^2 + k_2(j\omega) + 1\}$$

The quadratic form can of course occur in the denominator or the numerator of a transfer function (see Figure 5.6). To deal with this function we need to have a greater exposure to damping ratios and other terms, so we will not pursue this just at the moment.

Also, we can easily obtain **gain and phase margins** from Bode diagrams and the following example will show this.

❏ *Example 5.1*

If we take the following transfer function

$$GH(j\omega) = \frac{1.43}{j\omega(1+0.5j\omega)(1+0.01j\omega)}$$

then we can use the recognition of standard form functions and plot each part separately.

1. The gain of the transfer function (k) is 1.43, this has no phase and so will have the same magnitude regardless of angular frequency,

$$20 \log_{10}|1.43| = 3.1 \text{ dB}.$$

Hence, we can put a line at 3.1 dB across the width of the plot.

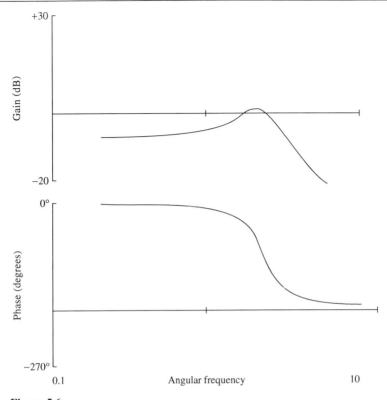

Figure 5.6

2. $1/j\omega$ will have a modulus which will decrease as ω increases. We have seen previously that this function will have a value of 40 dB at $\omega = 0.01$ rad/s and will fall off at −20 dB/decade.
3. $1/(1 + 0.5j\omega)$ will have a value of 0 dB until the break frequency is reached ($1/0.5 = 2$ rad/s) and then it will fall off at −20 dB/decade.
4. $1/(1 + 0.01j\omega)$ will also have a value of 0 dB until it reaches its corner or break frequency ($1/0.01 = 100$ rad/s) and then it too will fall at −20 dB/decade.

When we add these asymptotes up we get a function which starts at approximately 43 dB at the lowest frequency (0.01 rad/s). The characteristic then falls off at −20 dB/decade until the first corner frequency of 2 rad/s then falls off at −40 dB/decade. When the second corner frequency is reached (100 rad/s) the characteristic drops at the rate of −60 dB/decade.

Plotting the phase can be carried out in a similar manner

1. 1.43 has no phase.
2. $1/j\omega$ starts at zero frequency at −90° and stays at −90° until infinity.
3. $1/(1 + 0.5j\omega)$ varies in phase from zero at zero frequency to −90° at infinity but it is important to remember that the phase will be −45° at the break frequency.
4. $1/(1 + 0.01j\omega)$ also varies from zero phase at zero frequency to −90° at infinity, and again the phase will be −45° at 100 rad/s.

The phase plots must then be added together to plot the overall phase characteristic (see Figures 5.7a and b on pp. 46–47).

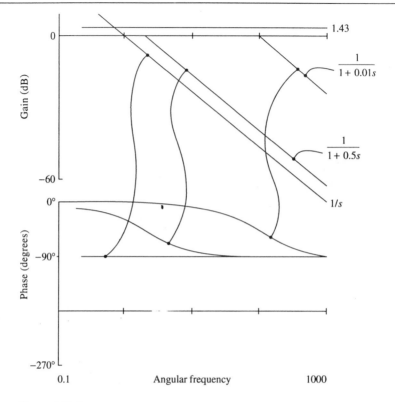

Figure 5.7(a)

(a) To determine the gain margin, we need to locate the point at which the phase characteristic crosses the $-180°$ axis. At this value of angular frequency, if the gain is negative, then the system is stable, if it is positive then the system is unstable.

(b) To calculate the phase margin we need the system phase when its gain is 0 dB. The difference between this and 180° is the margin.

In the example shown in Figure 5.7c the phase margin is $+52°$ and the gain margin is $+37$ dB. If a system has both gain margin and phase margin of zero (unity) then the system is 'marginally stable', which in practical terms is unstable.

As a rule of thumb a gain margin of between 8 and 20 dB and a phase margin of between 30° and 60° will assure stability. The problem is that if a decrease in gain is necessary to make a system stable it also has the effect of increasing the steady-state velocity error.

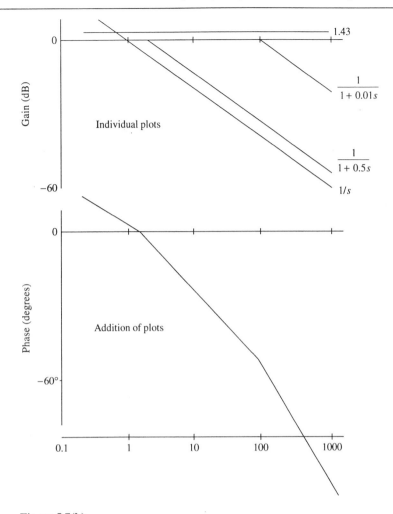

Figure 5.7(b)

Determination of a Transfer Function from a Bode plot

This is simply a reversal of the previous process. If we do not see a $j\omega$ or $1/j\omega$ term then the gain at the origin will give us the Transfer Function gain constant (k). Remembering that if

$$20 \log_{10}|k| = x \text{ then } k = 10^{x/20}$$

for example, if

$$k = 100 \text{ then } 20 \log_{10}|100| = 40$$

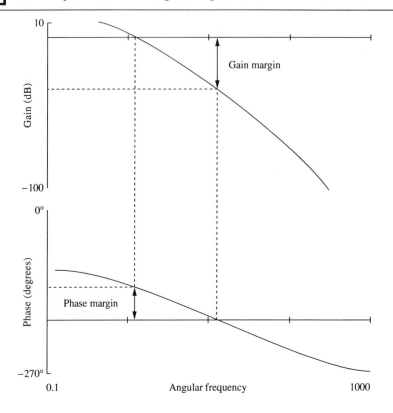

Figure 5.7(c)

and

$$10^{40/20} = 10^2 = 100$$

then we must determine the break points remembering that the break frequency = $1/\tau$, where τ is the time constant and also remembering that

$$1 + j\omega\tau \text{ has a slope of } +20 \text{ dB/decade}$$

$$1/(1 + j\omega\tau) \text{ has a slope of } -20 \text{ dB/decade}$$

and also

$$(1 + j\omega\tau)n \text{ has a slope of } +20n \text{ dB/decade}$$

$$1/(1 + j\omega\tau)n \text{ has a slope of } -20 \text{ dB/decade}.$$

❑ *Example 5.2*

Use straight line approximation and sketch a Bode plot for the following transfer functions. Estimate both gain and phase margins from the diagrams

(1) $Gp(s) = \dfrac{3}{s(1+0.125s)(1+0.25s)}$

(2) $\dfrac{10}{s(0.05s+1)(1+0.1s)}.$

Appendix

The solutions to Example 5.2 are given in Figures 5.8 and 5.9.

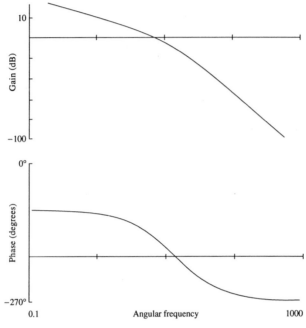

Figure 5.8

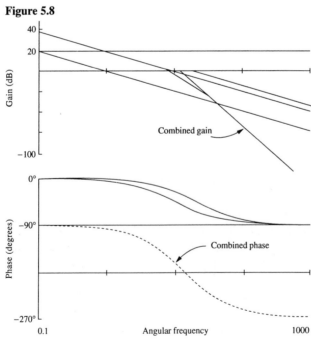

Figure 5.9

Practical task: open loop test on a Wein bridge oscillator

You will need the following equipment

- oscilloscope
- signal generator
- digital voltmeter
- phase meter
- proto-board
- variable resistor (R_f) for range 50–100 kΩ, a decade box would be ideal
- resistors: R_i = 39 kΩ, R_1 and R_2 = 10 kΩ
- capacitors: C_1 and C_2 = 10 nF
- 741 operational amplifier
- ± 15 V, centre zero, power supply.

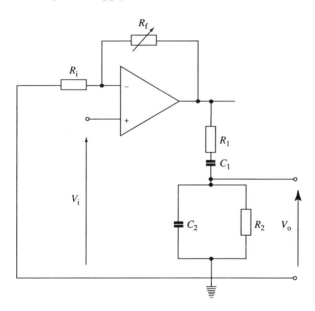

Figure 5.10

1. Contruct the circuit as shown in the diagram.
2. Set the signal generator to 3 V peak (V_i).
3. Vary the frequency from 160 Hz to 16 kHz in suitable steps.
4. Record V_o in magnitude and in phase angle with reference to V_i.
5. Tabulate results for R_f = 56 kΩ, 78 kΩ and 100 kΩ. Calculate and record the amplitude ratio V_o/V_i in dB and the relative phae shift for each step and for each value of R_f.
6. Plot the results onto log/linear graph paper as a Bode plot.
7. Use the plots of open loop data in conjunction with block diagram and frequency response theory to illustrate the stability criterion. Remember that the Wein network will give a feedback fraction of 1/3.

6

Compensation

We have spent some time in Chapters 4 and 5 looking at ways of determining stability and adjustment margins. What we need to do now is look at methods for compensating for instability or to adjust characteristics that would be otherwise undesirable. We can adjust the output characteristic of a system by adjusting the input signal or the system gain. In addition to this we can add components with the sole purpose of compensation. To fully cover the topic of compensation would be beyond the scope of this book; although readers interested in this topic can find many textbooks covering the subject.

We will now examine a few examples to acquire an insight into the compensation process and its philosophy.

Gain factor (K_0) compensation

We have seen that a stable system will have a gain margin which shows us the factor by which we can increase gain before the system becomes unstable. It follows, and is demonstrated by Figure 6.1, that a system which is unstable can generally be stabilised by reducing its gain.

The difficulty is that the reduction in gain will reduce the final steady-state value to be reached, but this will still occur in the same time frame as before. Hence, the rate of increase will be slower, and this slower response is often of great importance (see Figure 6.2).

Let us look at an example of gain factor reduction.

❑ *Example 6.1*

Data from an open-loop Frequency Response test are given overleaf in Table 6.1. Using the data a Bode plot is drawn which can be used to determine whether the system is stable. We can then determine the reduction of the system gain required to give a 10 dB gain margin. The Bode plots shown in Figure 6.3 have been plotted using the data from Table 6.1.

Consider first the curve (a) which is a plot of the uncompensated function. If we find the point on the gain curve which coincides with −180°, we find that we must measure downwards from the gain curve to the 0 dB axis. This shows that the system is unstable, in fact the gain at −180° is 17.54 dB (7.53). In order to achieve the desired gain margin (10 dB) a gain of 0.31623 at −180° is needed. This means that the curve must move downward to the position shown in (b).

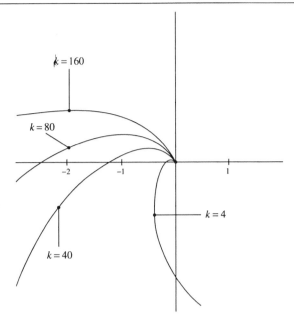

Figure 6.1

To do this a gain factor (K_0) reduction of

$$\frac{0.316227}{7.53} = 0.042$$

is needed. For example, if the original gain was 40 then the new gain is 1.68.

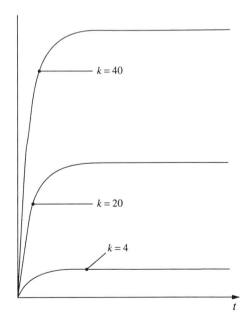

Figure 6.2

Table 6.1

ω (rad/s)	Gain (dB)	Phase (degrees)
0.2	43	−137.3°
0.4	32.7	−158.5°
0.8	21.62	−137.3°
1.232	14.07	−175°
1.69	20.43	−184.6°
2.23	3.326	−79.6°
3.1	−2.74	−198.9°
4.0	−8.334	−216°
5.0	−12.52	−222.1°
6.0	−16.88	−228.2°
7.0	−20.5	−232.9°
8.2	−24.18	−237.3°
9.25	−27.02	−240.3°
10.0	−28.93	−242.3°

It is easy to see that we can achieve a more stable gain and/or phase margin by adjusting the system gain, but this does bring other problems and so we will look at alternative methods.

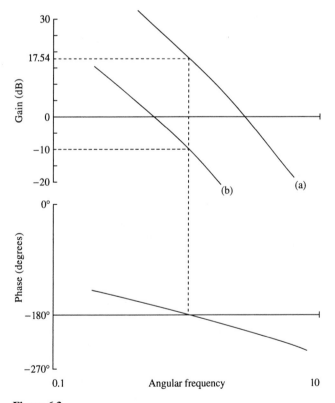

Figure 6.3

The Phase Lead network

This is one series compensation network, comprised of resistors and a capacitors, which is fairly easy to analyse and will help us to observe the compensation philosophy in operation.

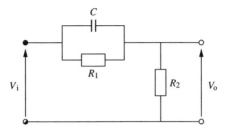

To derive the Transfer Function of such a circuit we will put $s = j\omega$ and first find an equivalent impedance to the parallel network of R_1 and $1/sC$, then use the voltage divider to find V_o/V_i.

1. Using product over sum

$$\frac{R_1 \cdot \dfrac{1}{sC}}{R_1 + \dfrac{1}{sC}} = \frac{\dfrac{R_1}{sC}}{\dfrac{sCR_1+1}{sC}} = \frac{R_1}{1+sCR_1}.$$

2. Now we can use the voltage divider for

$$\frac{V_o}{V_i} = \frac{R_2}{R_2 + \left(\dfrac{R_1}{1+sCR_1}\right)}.$$

Multiply top and bottom by $(1 + sCR_1)$

$$= \frac{R_2(1+sCR_1)}{R_2(1+sCR_1)+R_1}$$

$$= \frac{R_2}{R_1+R_2} \cdot \frac{sCR_1+1}{s\left(\dfrac{R_1R_2}{R_1+R_2}\right)C+1}.$$

If we put

$$\tau_1 = R_1C \quad \text{and} \quad \tau_2 = \left(\frac{R_1R_2}{R_1+R_2}\right)C$$

then

$$\frac{V_o}{V_i} = \frac{R_2}{R_1 + R_2} \cdot \frac{1 + s\tau_1}{1 + s\tau_2}.$$

The voltage divider is often referred to as α and the Transfer Function can be given as

$$\frac{V_o}{V_i} = \frac{\alpha(1 + s\tau)}{1 + s\alpha\tau}.$$

As the denominator is greater than the numerator, then the Transfer Function is smaller than unity for all frequencies, attenuating the input signal. As well as this, the circuit will have two corner frequencies given by the reciprocals of τ_1 and τ_2. The frequency response of the network is shown in Figure 6.4.

To examine the frequency response we must remember that we will be adding this waveform or characteristic to that of the plant to adapt the output.

The behaviour of the phase lead circuit can be best examined in three areas:

1. At a very low angular frequency $(<\omega_L)$ the capacitive reactance of the capacitor C is very high and will appear as an a.c. open circuit. Consequently, the output voltage V_o is attenuated by the voltage divider

$$\frac{R_2}{R_1 + R_2}$$

giving a negative dB output at low frequency.

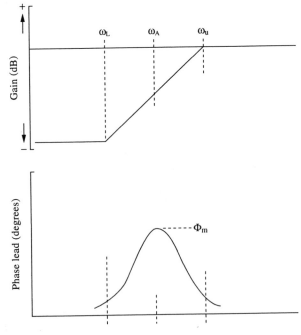

Figure 6.4

2. At angular frequencies above the upper cut-off value of ω_u, the capacitive reactance is small and the output voltage V_o will be (very nearly) equal to the input voltage V_i and hence a gain of 0 dB and phase shift of virtually 0° is seen.
3. Between ω_L and ω_u the capacitive reactance will have a finite value giving an output to input gain ratio between zero and $-x$ dB.

The phase shift between V_o and V_i will lead for values between zero and Φ_m. Φ_m occurs at the geometric mean of ω_L and ω_u shown as ω_A, i.e. $\omega_A = \sqrt{(\omega_L * \omega_u)}$. Also it can be shown that

$$\Phi_m = \tan^{-1}\frac{1-\alpha}{2\sqrt{\alpha}},$$

where α is the fraction of the voltage divider

$$\frac{R_2}{R_1 + R_2}.$$

It is often useful to be able to calculate the ratio α for circuit design from the value of Φ_m or to determine the ratio between R_1 and R_2. In order to do this we need to be able to transpose the equation for Φ_m: this will be easier if it is in another form.

Remember that if $\tan \Phi_m$ is given by $(1 - \alpha)/(2\sqrt{\alpha})$ and this can be modelled as a right-angled triangle (see the diagram below) with opposite side $= 1 - \alpha$ and adjacent side $= 2\sqrt{\alpha}$. From this we can determine that the hypotenuse has a value of $1 + \alpha$ so that

$$\sin \Phi_m = \frac{1-\alpha}{1+\alpha}$$

and

$$\alpha = \frac{1-\sin \Phi_m}{1+\sin \Phi_m}.$$

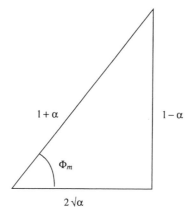

Effect of series Phase Lead circuit

Consider the following example: a system has an open-loop Transfer Function

$$Gp(s) = \frac{40}{(1+0.2s)(1+0.3s)(1+0.4s)}.$$

Constructing a Nyquist diagram will show that this is an unstable function (see Figure 6.5).

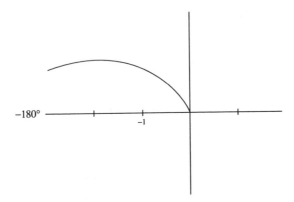

Figure 6.5

We could stabilise the system by reducing the gain but, as we have seen, this would have undesirable effects on the system response. Consequently, we will construct a Phase Lead circuit; this is shown in Figure 6.6. This will have a Transfer Function of

$$\frac{0.31973(1+0.1s)}{1+0.031973s}.$$

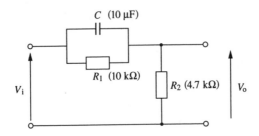

Figure 6.6

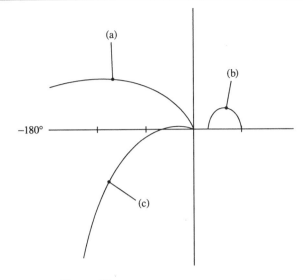

Figure 6.7

Consider the Nyquist diagram shown in Figure 6.7. Curve (a) is the unstable characteristic of 6.5 V, (b) is the Nyquist diagram of the Phase Lead circuit and (c) is the multiple of these.

At $\omega = 10.47$ rad/s: (a) has a value of 1.208 and an angle of $-213.5°$, (b) has a value of 0.439 and an angle of $27.83°$ and (c) is the multiple of (a) and (b) and has a value of (1.208) (0.439) and an angle of $(-213.5° + 27.83°) = 0.5303$ and an angle of $185.67°$.

Note that this example is designed to show the principle of compensation only with no other consideration.

The Phase Lag circuit

The phase lag compensator will introduce attenuation and hence stability, but only in the frequency region where the stability margin is defined. This means that low frequency characteristics will be largely unaffected but dynamic performance will be improved. This chapter will show that the lag compensator may not always improve performance and there are some systems in which it has no effect at all.

Consider the phase lag circuit shown in Figure 6.8 which comprises two resistors and one capacitor.

As we have done previously we can substitute s for $j\omega$ and derive a transfer function

$$\frac{V_o}{V_i} = \frac{R_2 + \dfrac{1}{sC}}{R_1 + R_2 + \dfrac{1}{sC}}.$$

Multiplying the top and bottom by sC gives

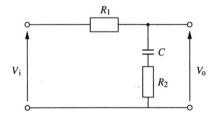

Figure 6.8

$$\frac{V_o}{V_i} = \frac{1 + sCR_2}{1 + sC(R_1 + R_2)}.$$

If we put $\tau_1 = R_2C$ and $\tau_2 = C(R_1 + R_2)$ then

$$\frac{V_o}{V_i} = \frac{1 + \tau_1 s}{1 + s\tau_2}.$$

The Bode plots for this circuit are shown in Figure 6.9.

To examine this frequency response, we can divide the plots into three regions as we did with the phase lead circuit.

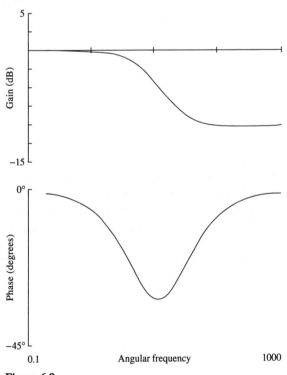

Figure 6.9

1. For very low values of angular velocity ($<\omega_L$) the reactance of the capacitor is so high as to appear as an open circuit. This ensures very little current is drawn and output voltage is very close to input voltage. The effect of this is to give a transfer function of very nearly one (0 dB) with very little phase difference between input voltage and output voltage.

2. At values of angular velocity above the upper cut-off value ω_u the capacitor will have a very low reactance and hence the output voltage will be given (approximately) by the voltage divider fraction and this will be less then one. Hence, the gain will be a negative decibel ratio with a negligible phase angle.

3. For values of angular velocity between ω_L and ω_u there will be a gradual decrease in gain with phase angle varying from nearly zero through a maximum phase shift (Φ_m) back to nearly zero.

Maximum phase lag occurs at the geometric mean of $\omega_L\omega_u$ which we will call ω_A

$$\omega_A = \sqrt{(\omega_L\omega_u)}.$$

We will examine the effect of inserting a phase lag compensator in series with the plant from the example used in the section on phase lead compensation (page 58) (see Figure 6.10).

We see that the addition of the arbitrary circuit actually increases the system instability. This is due, mainly, to the fact that it is phase lag in the system which causes instability and here we are adding still more phase lag. We can use this type of compensator, and indeed it may have its merits, but it must have an upper cut-off frequency ω_u which is below the gain cross-over frequency of the system. Our unstable system has a gain cross-over frequency of about 11.4 rad/s and we must use this parameter in our compensator design.

In order to design this compensator we would need to know the desired phase margin after compensation. Let us say that we require 40° and we must allow for 5°

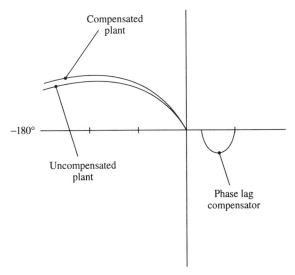

Figure 6.10

which is the approximate phase lag introduced by the compensator circuit

$$180 - 40 - 5 = 135°.$$

We now need to know the angular velocity at which the original system exhibits this phase lag. This occurs at about 3.5 rad/s (ω_d) and at that frequency the system has a gain of 13.37 (22.52 dB).

Using the general rules

$$\omega_u = 0.1\omega_d \text{ (design frequency)}$$

and

$$\omega_L = \omega_u/(\text{gain at } \omega_d).$$

Then

$$\omega_u = 0.35 \text{ and } \omega_L = 0.026$$

giving

$$Gc(s) = \frac{1 + \dfrac{s}{0.35}}{1 + \dfrac{s}{0.026}}.$$

Figure 6.11 shows the effect of this compensator in series with the plant. Obviously having obtained the compensator transfer function, we would need to determine the nominal value of its discrete components. However, we will not do this here as we are only looking at the general philosophy of compensator design. Any further reading

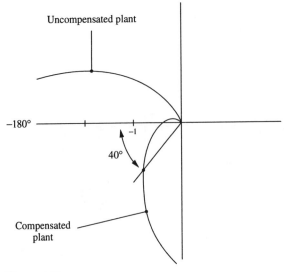

Figure 6.11

on this subject should include double-lead compensators as well as internal or parallel compensators and acceleration feedback compensation.

Let us conclude here with a summary of the salient points in this chapter:

■ **Phase lead compensators** increase system bandwidth and consequently the speed of response of the closed-loop system. However, care must be taken not to increase the bandwidth excessively, making the system overly responsive and susceptible to noise. Systems designed and stabilised using linear models may not be possible to achieve due to the limitations of the input signal. Phase lead is the only way to stabilise a system which is unstable or marginally stable in the open-loop state.

■ **Phase lag compensators** reduce the speed of response and this can itself introduce additional problems if they are not designed with care.

■ **Lag lead compensators** utilise both characteristics and are used where a bandwidth cannot be achieved or maintained by one or the other alone.

■ **Double lead compensators** are used where sufficient phase lead cannot be achieved by a phase lead device alone.

Practical task: the Phase Lead circuit

You will need the following equipment:

- oscilloscope
- signal generator
- digital voltmeter
- phasemeter
- proto-board
- resistors: $R_1 = 22\ k\Omega$ (and 47 kΩ), $R_2 = 10\ k\Omega$
- capacitor: $C = 0.1\ \mu F.$

The circuit diagram is as here:

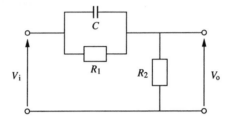

Method

1. Construct the circuit as shown in the diagram.
2. Set the signal generator to 3 V peak (V_i).
3. Vary the frequency from 10 Hz to 1 kHz in suitable steps, e.g. 25–200 Hz in 25 Hz steps, then in 50 Hz steps to 500 Hz and finally in 100 Hz steps.
4. Record V_o and its phase angle with relation to V_i per step.
5. Tabulate your results, with the amplitude ratio V_o/V_i in polar form.
6. Plot your results on linear axes as a Nyquist diagram.
7. Evaluate Φ_{max} and ω_{max} and compare these to those determined by calculation.
8. Repeat the above procedure with $R_1 = 47\ k\Omega$.

7 Introduction to Computer methods

CODAS (**CO**ntrol **D**esign **A**nd **S**imulation) is an interactive suite of programs which assist in giving the student a clear feel for a subject which can be otherwise difficult to conceptualise. This program has been used by the author in this book.

We shall be considering this program several times in this book but at this point let us use it to illustrate some of the frequency response and compensation principles. The assumption is that the reader has no prior knowledge of, or experience with, CODAS.

Getting started

At the DOS prompt type 'CODAS' and press 'return'. You may need to change the drive if you are working from a floppy disk. Once the program is loaded it will instruct you to press any key and this takes you into the time domain screen. Although we are currently investigating the frequency domain let us look at this screen as an introduction.

Type 'L' and when asked for the name of the file to load type 'Example'. You will now be given a choice of three selections:

- System
- Environment
- Both.

Choose 'System' and once the return is pressed a Transfer Function will appear in the bottom left-hand corner of the screen.

Time Domain

The Time Domain environment will show the response of the function $\{Gp(s)\}$ to a unit step input (see Figure 7.1). As the environment title suggests this response will be as a function of time and not of frequency (or anything else).

The input signal can be changed from a unit step; this will be discussed later.

It may well be that the default settings for time and gain may not suit the case and we may need to change these. However, firstly the default setting in this environment is the closed loop and even though we have been examining the open loop behaviour we will use this initially.

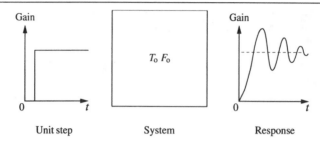

Unit step   System   Response

Figure 7.1

Press 'G' (for 'go') and see the closed loop Time Domain response to a unit step. We see that we need to change the scale settings of the axes. Select 'Y' and for minimum we will enter zero and for maximum 0.75 (return). Then for 'X' we will enter zero minimum and 25 maximum (return). When we press 'G' we will see a full picture of an underdamped oscillation (see the chapter on Time Response, page 90).

Select 'O' for open loop and press 'G' and we see that the signal is quite different. The axes must be reset for an accurate picture.

Set Y for minimum 0 and maximum 1.5 and X for mimimum 0 and maximum 5. Press 'G' again and we see a full picture of this response.

Frequency response: the Nyquist diagram

We will now turn to frequency response as this is a topic we have looked at earlier on in the book.

Select F8 and the screen will change to the Frequency Domain and to the Nyquist diagram. (If the Nyquist diagram does not appear when the action below is taken select 'V' for view and select option 1.)

We still have the Transfer Function of the example on the screen and if 'G' is pressed we will see a Nyquist diagram for the function of values of angular velocity between 0.2 and 8 rad/s. The range of angular velocity is shown on the screen as frequency and we can change the frequency range by selecting 'f' and entering minimum and maximum values needed. However, we will examine the characteristic currently shown on the screen first. If you press the '?' key a cursor will appear on the screen, and this can be moved along the characteristic by using the right and left shift controls.

Use the right and left shift keys to move the cursor along the characteristic and each time you allow it to stop there will be a data display at the foot of the screen. This display will indicate the angular velocity which corresponds to the current coordinates and will give information on the vector at that point. If we take the cursor to the $-180°$ axis we can see the gain at this point which will enable us to determine the gain margin.

Without changing the number of points calculated and used to plot the diagram we can only get $-179.6°$ and at this point we see that gain is 0.5131. Press '=' and a calculator appears on the screen. We want 1/0.5131 and we see that this gives a gain margin of 1.9489. We cannot evaluate phase margin in this case since the characteristic does not cross the unit circle.

Remove the cursor by selecting 'escape'. If you still have the calculator on the screen you will need to select 'escape' twice.

The system gain constant can be varied by the 'k' function. If you select 'k' and put in a value of 1.9489 then 'return'. You will now have a system gain which will be critically stable. Press 'G' and the characteristic will be redrawn and we will see that it passes through the −1 point as it crosses the −180° axis.

The Bode plot

Select 'V'; six alternatives will be offered. Select 6, 'return' and you will see Bode gain and phase plot axes. Press 'G' and the gain and phase characteristics will be drawn. Enable the cursor and bring to −179.6° and we see the gain is 0.5131 as before.

Turn now to Figure 5.5 (page 40) and enter the following Transfer Function

$$Gp(s) = \frac{1.43}{s(1+0.5s)(1+0.01s)}.$$

Do this by first selecting 'N' and then entering the numerator 1.43. After 'return' select 'D' and then enter the denominator $s(1 + 0.5s) (1 + 0.01s)$.

The screen and the parameters can be set to get accurate data from the plot. To set the gain range select 'V' and Bode gain. Once you are in this screen you can select 'Y' and insert minimum = −100 and maximum = 80. As we are using default phase limits we can return to Bode gain and phase. To do this select 'V' and 6. Set frequency range, select 'f', minimum 0.001 and maximum 1000 rad/s. Finally change the number of points calculated and plotted. Select 'j' and insert 1000 and 'return'. Now press 'G' and the characteristic is drawn on the screen. Enable the cursor (?) and find the gain at −180°. This is 0.01404 giving a gain margin of 71.225. Find the point where the characteristic crosses the 0 dB axis and a phase margin of about 58° is seen.

Compensation

This section briefly examines gain constant compensation, phase lead circuit and phase lag circuit. These will be reviewed below in turn using CODAS.

Gain constant compensation

❑ Example 7.1

Take the following function

$$Gp(s) = \frac{45}{(s+1)(s^2+1.4s+15)}.$$

1. Is the system represented by this function stable?
2. If it is unstable, by what factor will we need to decrease the gain in order for the system to have a gain margin of 6 dB?
3. We could investigate this problem using a Nyquist diagram or a Bode plot. However, we shall use the Bode plot as it may make it easier to see what is achieved and how we would achieve it. Let us first produce a Bode plot of the function – this is shown in Figure 7.2.

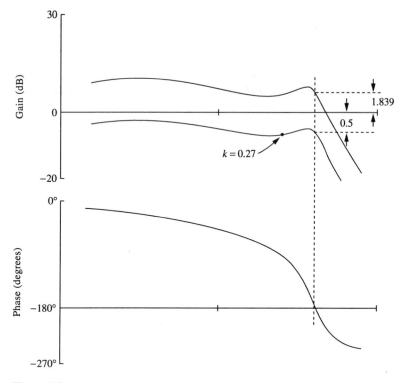

Figure 7.2

At the phase cross-over frequency, when the phase is –180°, the gain has a value of 1.851 (5.35 dB). This means that the system is unstable. Remember that a stable system has a gain of less than 1 at –180°, this would be a negative dB value. To find the gain margin the gain must fall from 1.851 to 0.5 at the –180° axis. This can be done by simply reducing gain. The gain needs to be reduced by

$$\frac{1.851}{0.5} = 0.27.$$

This value can be entered for the gain factor 'k' and the curve can be redrawn to see that it is below the 0 dB axis at –180°, hence it must now be stable.

Enable the cursor and take it to the point where the phase is –180° and we see that the gain at –179.8° is 0.4997 which is close enough for our purposes.

Phase lead compensation

Compensation has not been considered sufficiently to 'design' a network and so we will investigate the affect of inserting a phase lead network. Additionally CODAS will be used to confirm the equations introduced in the section on the phase lead network in Chapter 6 (page 55).

Enter the Transfer Function

$$Gp(s) = \frac{5}{s(1+5s)(1+1.5s)}.$$

If a Nyquist diagram is produced it can be seen that the system represented is unstable. In order to investigate the effect of inserting a series compensator let us consider the network represented by Figure 7.3.

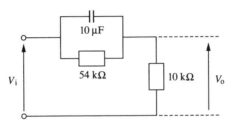

Figure 7.3

The network in Figure 7.3 has the following Transfer Function

$$Gc(s) = \frac{0.15625(1+0.54s)}{1+0.084375s}.$$

Using the F3/F4 function we can enter the compensator Transfer Function $Gc(s)$ and it enables us to flip from $Gp(s)$ to the product $Gp(s)Gc(s)$.

Figure 7.4 shows the effect of the compensator and also the Nyquist diagram of the phase lead network. To investigate the phase lead circuit it is necessary to enter its Transfer Function as $Gp(s)$ and change to Bode gain and phase. To obtain the information, extend the frequency range (say from 0.1 to 1000 rad/s) and the number of points plotted (to 1000 say). It is important to remember to extend the frequency range use 'f' and for the number of points 'j'.

Figure 7.5 is a Bode plot for the circuit shown in Figure 7.3 from which ω_L, ω_u, ω_A, Φ_m and hence α can be found. These component values exist and hence the values can be confirmed by calculation.

Readers can check these calculations against the results in the appendix.

Phase lag compensation

One of the main contributing factors to instability in control systems is the natural 'phase lag' which occurs due to inertia, friction, stiction and so on. All these increase as the system ages or requires maintenance; one would think that the last thing needed is to add further phase lag. However, there are some sound reasons for choosing phase lag compensation, but care must be taken to design it safely.

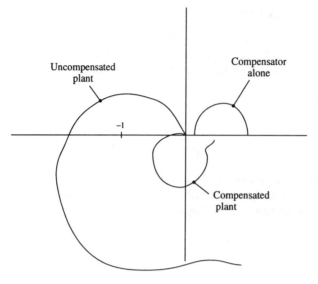

Figure 7.4

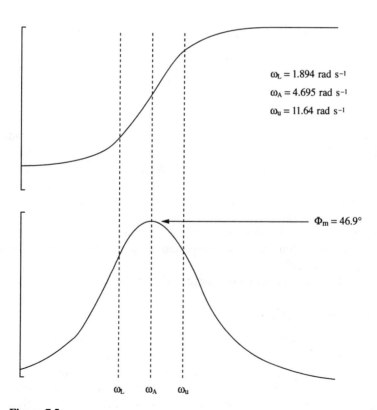

$\omega_L = 1.894$ rad s^{-1}
$\omega_A = 4.695$ rad s^{-1}
$\omega_u = 11.64$ rad s^{-1}

$\Phi_m = 46.9°$

ω_L ω_A ω_u

Figure 7.5

The main consideration is that the phase lag bulge must occur well below the gain cross-over frequency of the unstable system. We shall not attempt to design a compensator from first principles as this is outside the scope of this book but it is useful to investigate the principles.

Consider the example of a system with

$$Gp(s) = \frac{1000}{s(1+s)(16+s)}.$$

A Bode plot can be produced using CODAS and it can be seen that the system is unstable with a cross-over frequency of about 7.3 rad/s. The phase margin required needs to be determined and although it is not important use 45°, and after allow 5° for the compensator, this leaves

$$180 - 45 - 5 = 130°.$$

The angular velocity at which this phase occurs in the unstable system now needs to be determined. This will be the design frequency ω_d. The plot in Figure 7.6 shows that this occurs are $\omega = 0.7474$ rad/s when the system gain is 66.91.

Now $\omega_u = 0.1\omega_d$ and

$$\omega_L = \frac{\omega_u}{\text{gain at } \omega_d}$$

giving $\omega_u = 0.07474$ and $\omega_L = 1.12 \times 10^{-3}$ rad/s

$$Gc(s) = \frac{1 + \dfrac{s}{\omega_u}}{1 + \dfrac{s}{\omega_u}}.$$

When entered as $Gp(s)Gc(s)$ the following parameters are deduced: gain margin = 16.6 and phase margin = 45°.

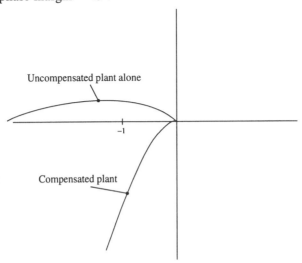

Uncompensated plant alone

−1

Compensated plant

Figure 7.6

Examples

Using CODAS, or otherwise, solve the following:

1. A Control System has the following open-loop Transfer Function

$$Gp(s) = \frac{20}{s(1+0.4s)(1+0.25s)}.$$

(a) Determine by how much the gain constant of the system must be reduced to give a gain margin of 8 dB and give the phase margin at this value.

(b) Insert a compensator into the feedforward path, the compensator will have a transfer function

$$Gp(s) = \frac{1+0.2s}{1+0.02s}.$$

What are the gain and phase margins now?

(c) Investigate the compensator of (b), is it phase lead or phase lag?

(d) Attempt to design a compensator of the same type as that in (b) but which will result in a phase margin of 20° without reduction of gain. Compare the response rate of the uncompensated system to that of the compensated one.

2. A d.c. generator with Ward–Leonard speed control is analysed and represented by Figure 7.7. Determine the gain and phase margins for this system and also the gain and phase cross-over frequencies.

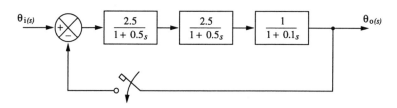

Figure 7.7

Appendix: Computer methods

The following is confirmation, by calculation, of the circuit parameters from Figure 7.3

$$\alpha = \frac{R_2}{R_1 + R_2} = \frac{10k}{10k + 54k} = 0.15625$$

$$\Phi_m = \sin^{-1}\frac{1-\alpha}{1+\alpha} = 46.9°$$

$$\tau = CR_1 = 0.054 \text{ s}$$

$$\omega_m = \frac{1}{\tau\sqrt{\alpha}} = 4.685 \text{ rad/s.}$$

❏ Example 7.2

1. From the Bode plot of this function it can be seen that there is a gain of about 3.07 at −180°. We want this gain to be 0.4, hence the gain needs to be reduced by a factor of

$$\frac{0.4}{3.07} = 0.13.$$

Enter $k = 0.13$ into the CODAS model and you will see that the gain is equal to 0.4 at −180° with a phase margin of 27°.
2. When $Gc(s)$ is inserted there is a gain margin of 2.139 and a phase margin of 8.4°.
3. By either entering $Gc(s)$ as $Gp(s)$ and plotting a graph or by putting $s = j\omega$ and then any value of ω and calculating gain and phase we see that this is phase lead device.
4. After some trial and error a Transfer Function can be formulated that achieves a phase margin of 20°. There will be many different transfer function ratios, but use the following

$$Gc(s) = \frac{1+0.38s}{1+0.0106s}.$$

In practice this would have to be achieved by more than one stage of compensator which should not exceed

$$Gc(s) = \frac{1+s\tau}{1+0.1s\tau}.$$

However, this compensator gives

$$\Phi_m \text{ of } 71.04° \text{ at } \omega_m = 15.66 \text{ rad/s.}$$

From this information we can calculate the resistor ratio α, since

$$\alpha = \frac{1-\sin\Phi_m}{1+\sin\Phi_m}.$$

In this case $\alpha = 0.028$ and, as already stated, we would not want α to be less than 0.172 in practice.

Now that the value of α has been determined it is possible to calculate τ because

$$\omega = \frac{1}{\tau\sqrt{\alpha}}$$

giving $\tau = 0.3824$.

We do not have sufficient information to design the circuit and so we will choose a component value and the other two components will have the correct ratio to that. Take R_2 as 4.7 kΩ which gives $R_1 = 163.157$ kΩ and $C = 2.344$ µF.

Figure 7.8 shows the system response to a unit step input.

❑ Example 7.3

The feedforward Transfer Function can be found from the multiple of the component Transfer Functions

$$GH(s) = \frac{6.25}{(1+0.5s)^2(1+0.1s)}$$

Enter this data into CODAS and produce either a Nyquist diagram or a Bode plot. By invoking the cursor the following information can be deduced: when gain = 1, phase is −154.1°, this gives a phase margin of $180 - 154.1 = 25.9°$; when the phase is −180° then the gain is 0.423 giving a gain margin of 2.364.

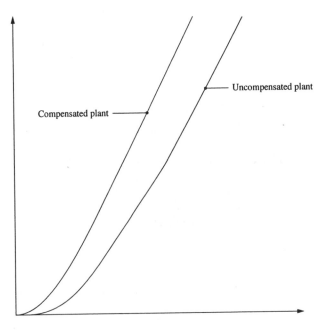

Compensated plant

Uncompensated plant

Figure 7.8

8 Control Engineering Mathematics

So far each chapter has looked at a different tool for building mathematical models of systems. Up to now only algebra and complex numbers were needed, however, it is not possible to progress further into the subject without forming and solving differential equations. In almost all of the systems considered variables will be functions of time and so it is necessary to form and solve differential equations.

Readers should have encountered differential equations in their mathematics studies and will be aware of the problems associated with their manipulation and solution. As a consequence of this difficulty, operator methods are normally applied to Control Engineering problems; there are several operators which could be used. The D operator is probably the simplest, but there are problems in systems with non-zero initial conditions. Also, modern control software is all written in terms of the s operator or the Laplace transform.

In this chapter the Laplace transform will be reviewed with some attention to its derivation from first principles. However, it is not often necessary for the engineer to do this, rather, to use the transform as a tool.

The Laplace transform is defined by

$$L(f(t)) = \int_0^\infty f(t)e^{-st}dt = F(s).$$

Derivation of Laplace transforms

Derive the Laplace transform for $f(t) = K$, where K is a constant

$$L(f(t)) = \int_0^\infty f(t)e^{-st}$$

re-organising for the function K gives

$$L(f(t)) = \int_0^\infty Ke^{-st}\ dt$$

hence

$$L(f(t)) = \left[\frac{K}{-s} e^{-st} \right]_0^\infty$$

$$= \left[\frac{K}{-s} e^{-\infty} - \frac{K}{-s} e^0 \right]$$

$$L(f(t)) = F(s) = \frac{K}{s}.$$

It is possible to go on and derive all the Laplace transforms required, but it would eventually become necessary to use integration by parts. At this point in the book we only want to consider Laplace transforms so we will not proceed further. Table 8.1 lists all the transforms that we will need for the moment.

Table 8.1 Laplace transforms

		$f(t)$ from $t = 0$	$F(s) = L[f(t)]$
1.		unit impulse $s(1)$	1
2.		unit step $u(1)$	$\dfrac{1}{s}$
3.		ramp t	$\dfrac{1}{s^2}$
4.		nth order ramp t^n	$\dfrac{n!}{s^{n+1}}$
5.		exponentials $e^{-\alpha t}$	$\dfrac{1}{s+\alpha}$
6.		$1 - e^{-\alpha t}$	$\dfrac{\alpha}{s(s+\alpha)}$

Table 8.1 (*continued*)

	f(t) from t = 0	F(s) = L[F(t)]
7.	$te^{-\alpha t}$	$\dfrac{1}{(s+\alpha)^2}$
8.	$t^n e^{-\alpha t}$	$\dfrac{n!}{(s+\alpha)^{n+1}}$
9.	$e^{-\alpha t} - te^{-\beta t}$	$\dfrac{\beta+\alpha}{(s+\alpha)(s+\beta)}$
10.	sine wave $\sin \omega t$	$\dfrac{\omega}{s^2+\omega^2}$
11.	$\sin(\omega t + \phi)$	$\dfrac{\omega\cos\phi + s\sin\phi}{s^2+\omega^2}$
12.	$t\sin \omega t$	$\dfrac{2\omega s}{(s^2+\omega^2)^2}$
13.	cosine wave $\cos \omega t$	$\dfrac{s}{s^2+\omega^2}$
14.	$\cos(\omega t + \phi)$	$\dfrac{s\cos\phi - \omega\sin\phi}{s^2+\omega^2}$

Table 8.1 (*continued*)

	$f(t)$ from $t = 0$	$F(s) = L[f(t)]$
15.	$t \cos \omega t$	$\dfrac{s^2 - \omega^2}{(s^2 + \omega^2)^2}$
16.	$1 - \cos \omega t$	$\dfrac{\omega^2}{s(s^2 + \omega^2)^2}$
17.	$\sin \omega t - \cot \cos \omega t$	$\dfrac{2\omega^3}{(s^2 + \omega^2)^2}$
18.	$e^{-\alpha t} \cdot \sin \omega t$	$\dfrac{\omega}{(s+\alpha)^2 + \omega^2}$
19.	$e^{-\alpha t} \cdot \cos \omega t$	$\dfrac{(s+\alpha)}{(s+\alpha)^2 + \omega^2}$
20	$e^{-\alpha t}[\sin \omega t - \omega t \cos \omega t]$	$\dfrac{2\omega^3}{[(s+\alpha)^2 + \omega^2]}$
21.	$\sinh \omega t$	$\dfrac{\omega}{s^2 - \omega^2}$
22.	$\cosh \omega t$	$\dfrac{s}{s^2 - \omega^2}$

The first shift theorem tells us to replace s by $s - a$ for a multiple of e^{at}.

$t^2 \cdot e^{at}$

$\dfrac{2}{(s-a)^3}$

Remember

$$L(t^2) = \dfrac{n!}{s^{n+1}}$$

$$= \dfrac{2}{s^3}$$

Inverse Laplace transforms

After some practice readers will see that the process of using Laplace transforms is not too difficult. In most cases it is necessary to carry out some analytical process, then determine the Time Domain function that this would represent. Not surprisingly this final stage is called the inverse transform and is largely a matter of using tables. However, it is often first necessary to divide a compound expression into its partial fractions in order to be able to do this final stage.

❑ Example 8.1

Determine

$$L^{-1}\left[\frac{2s-6}{(s^2-6s+8)}\right].$$

As the denominator is of a higher degree than the numerator it is possible to factorise the quadratic denominator and then use partial fraction techniques and finally to use the tables to find $f(t)$, i.e.

$$\frac{2s-6}{(s-2)(s-4)}=\frac{1}{s-2}+\frac{1}{s-4}$$

which gives $f(t) = e^{2t} + e^{4t}$ by using the transform tables.

Theorems and initial conditions

The first shift theorem allows us to put $(s - a)$ when we see e^{at}. For example, if we want to find the Laplace transform of te^{4t}, we know from a pair number 3 that the transform of t is $1/s^2$ and so the shift theorem can be used to put $(s - 4)$ in place of s. Therefore

$$L(te^{4t}) = \frac{1}{(s-4)^2}.$$

The second shift theorem is used with discontinuous functions such as impulses – these will not be considered here.

Initial conditions

To make the mathematics easier it is often assumed that the initial conditions are zero when we energise, or are first interested in, the system. However, this is often not the case. For instance, in electrical circuits, capacitors and inductors store energy and dissipate it after the circuit is switched off – obviously this cannot be considered a zero initial condition. This non-zero initial condition is at its worst in the case of a switching circuit, solid state or otherwise. So, in Control Engineering it is necessary to be able to enter the initial conditions into the Laplace transformed equation.

The solution of differential equations

There are many techniques for solving differential equations. Here a few of the classical techniques will be reviewed and compared with the Laplace transform method.

By the classical method

Equations of the type

$$\frac{dy}{dt} = ky$$

have a solution of the form

$$y = Ae^{kt}$$

❑ Example 8.1

Solve the equation

$$\frac{d\theta}{dt} = 6\theta$$

given that $\theta = 3$ when $t = 0$. We can see that this equation is of the dy/dt form and so we can say that $\theta = Ae^{6t}$ and we can use initial conditions to evaluate A. Therefore

$$3 = Ae^{6(0)},$$

$$A = 3.$$

The particular solution is

$$\theta = 3e^{6t}.$$

By Laplace transformation

$$L\left\{ \frac{d\theta}{dt} = 6\theta \right\}$$

$\theta = 3$ when $t = 0$

$$s\theta_{(s)} - 3 = 6\theta_{(s)}$$

$$\theta_{(s)}(s-6) = 3$$

$$\theta_{(s)} = \frac{3}{s-6}$$

$$L^{-1} = \left[\frac{3}{s-6}\right]$$

$$\theta = 3e^{6t}.$$

By direct integration

If the differential equation can be arranged into the form $dy/dx = f(x)$ then both sides can be integrated to give a solution.

❑ *Example 8.2*

Solve the equation

$$\frac{d\theta}{dt} = 3t^2 - 6t + 5$$

$\theta = 2$ when $t = 0$.

Both sides can be integrated with respect to the variable t

$$\int \frac{d\theta}{dt} = \int 3t^2 - 6t + 5.$$

Giving us $\theta = t^3 - 3t^2 + 5t + C$ and we can use the initial conditions to find a value for C, $C = 2$ therefore the particular solution is $\theta = t^3 - 3t^2 + 5t + 2$.

By Laplace transformation

$$L\left\{\frac{d\theta}{dt} = 3t^2 - 6t + 5\right\}$$

$\theta = 2$ when $t = 0$

$$s\theta_{(s)} - 2 = \frac{6}{s^3} - \frac{6}{s^2} + \frac{5}{s}.$$

Rearrange for

$$\theta_{(s)} = \frac{6}{s^4} - \frac{6}{s^3} + \frac{5}{s^2} + \frac{2}{s}.$$

Find the inverse transform

$$L^{-1}\{\theta = t^3 - 3t^2 + 5t + 2\}.$$

Using the integrating factor (IF)

❑ **Example 8.3**

Solve

$$\frac{di}{dt} + 2i = 10e^{3t}, \quad i = 6A, \ t = 0$$

which is a mathematical model for the circuit current shown in Figure 8.1.

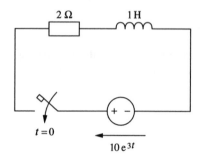

Figure 8.1

To solve this differential equation we will need to use an integrating factor. To find the IF we will compare the equation to

$$\frac{dy}{dx} + Py = Q$$

the IF is

$$\exp\left(\int P \, dx\right) = e^{2t}$$

and we multiply by this factor

$$\frac{di}{dt} e^{2t} + e^{2t}i = 10e^{5t}.$$

The left-hand side of this equation is the differential coefficient of

$$ie^{2t}$$

for example

$$\frac{d}{dt}\left[ie^{2t}\right] = 10e^{5t}.$$

and now we can integrate both sides

$$ie^{2t} = 2e^{5t} + C$$

and use initial conditions to evaluate C

$$i(t) = 2e^{3t} + 4e^{-2t}.$$

By Laplace transformation

$$L\left[\frac{di}{dt} + 2i = 10e^{3t}\right], \quad i = 6A \text{ at } t = 0$$

$$= si_{(s)} - 6 + 2i_{(s)} = \frac{10}{s-3}$$

rearrange for i (s)

$$i_{(s)} = \frac{6s-8}{(s-3)(s+2)}$$

and determine the partial fractions

$$\frac{6s-8}{(s-3)(s+2)} = \frac{2}{s-3} + \frac{4}{s+2}.$$

Finally, find the inverse Laplace transform

$$L^{-1}\left[\frac{2}{s-3} + \frac{4}{s+2}\right] = 2e^{3t} + 4e^{-2t}.$$

Examples of homogeneous differential equations solved by substitution and other techniques could also be included, but at the moment we are only interested in the s-operator method.

❏ Examples

Determine the following s-plane functions as functions of time:

1. $\dfrac{3s^2 - 7s}{(s-1)(s-2)(s-3)}$

2. $\dfrac{2s^2 - 4}{(s-3)(s-2)(s+1)}$

3. $\dfrac{3}{s}$

4. $\dfrac{5}{s+3}$

4. $\dfrac{5}{s+3}$

5. $\dfrac{4}{s^4}$

6. $\dfrac{s}{s^2+7}$

7. $\dfrac{3s-5}{s(s-2)(s+3)}$

8. $\dfrac{5s+1}{s^2-s-12}$

9. $\dfrac{20-s}{s^2+7s+12}.$

For (9) plot the time-domain function for $t = 0$ to 2.5 s in steps of 250 ms.

10. A Control System has a transfer function of

$$\dfrac{10}{s+5}$$

determine the output of the system as a function of time when a step input of 15 V is applied to the input terminals.

11. A series circuit has a resistance of 10 Ω, an inductance of 0.5 H and a capacitance of 500 μF. Take the circuit output as the capacitor voltage and determine the output voltage as a function of time when a voltage of 10 V d.c. is applied to the input terminals.

12. Determine the inverse Laplace transform of the following function

$$F(s) = \dfrac{2}{(s+3)(s-2)^2}.$$

Appendix: Control Engineering Mathematics

Worked solution to Example 8.1:

$$\frac{2s-6}{(s^2-6s+8)} = \frac{2s-6}{(s-2)(s-4)} = \frac{A}{(s-2)} + \frac{B}{(s-4)}.$$

Multiply out by the denominator or use the cover-up rule

$$2s - 6 = A\,(s-4) + B\,(s-2)$$

put $s = 4, B = 1$ and $s = 2, A = 1$

$$\frac{2s-6}{(s^2-6s+8)} = \frac{1}{(s-2)} + \frac{1}{(s-4)}$$

$$L^{-1}\left[\frac{1}{(s-2)} + \frac{1}{(s-4)}\right] = e^{2t} + e^{4t}.$$

❑ *Examples*

1. $$\frac{3s^2-7}{(s-1)(s-2)(s-3)} = \frac{A}{(s-1)} + \frac{B}{(s-2)} + \frac{C}{(s-3)}.$$

Cross multiply by the denominator, or use the cover-up rule

$$3s^2 - 7 = A\,(s{-}2)\,(s{-}3) + B\,(s{-}1)\,(s{-}3) + C\,(s{-}1)\,(s{-}2)$$

put $s = 1, A = -2;\ s = 2, B = -5$ and $s = 3, C = 10$.
Therefore

$$\frac{3s^2-7}{(s-1)(s-2)(s-3)} = \frac{10}{(s-3)} + \frac{5}{(s-2)} + \frac{2}{(s-1)}$$

$$L^{-1}\left[\frac{10}{(s-3)} - \frac{5}{(s-2)} - \frac{2}{(s-1)}\right] = 10e^{3t} - 5e^{2t} - 2e^{t}.$$

2. $$\frac{2s^2-4}{(s-3)(s-2)(s+1)} = \frac{A}{(s-3)} - \frac{B}{(s-2)} + \frac{C}{(s+1)}$$

multiply out or use the cover-up rule; put $s = 3, A = 14/4;\ s = 2, B = -(4/3)$ and $s = -1, C = -(1/6)$

$$\frac{2s^2 - 4}{(s-3)(s-2)(s+1)} = \frac{7}{2(s-3)} - \frac{4}{3(s-2)} + \frac{1}{6(s+1)}$$

$$L^{-1}\left[\frac{7}{2(s-3)} - \frac{4}{3(s-2)} - \frac{1}{6(s+1)}\right] = \frac{7}{2}e^{3t} - \frac{4}{3}e^{2t} - \frac{1}{6}e^{-t}.$$

3. Compare to pair number 2

$$L^{-1}\frac{3}{s} = 3.$$

4. Compare to pair number 5

$$L^{-1}\frac{5}{s+3} = 5e^{-3t}.$$

5. This is similar to pair number 4 with $n = 3$, but if this is the case it should be $6/s^4$ therefore there must be a constant multiplier which changes 6 to 4. If we put

$$k = \frac{3!}{s^{3+1}} = \frac{4}{s^4} \quad \text{then } k = \frac{2}{3}$$

i.e.

$$L^{-1}\left[\frac{s}{s^4}\right] = \frac{2}{3}t^3.$$

6. Compare to pair number 13

$$L^{-1}\left[\frac{s}{s^2 + 7}\right] = \cos\sqrt{7}t.$$

7. $$\frac{3s-5}{s(s-2)(s+3)} = \frac{A}{s} + \frac{B}{(s-2)} + \frac{C}{s+3}$$

multiply out by the denominator or use the cover-up rule, put $s = 0, A = 5/6$; $s = 2, B = 1/10$ and $s = -3, C = -14/15$

$$\frac{3s-5}{s(s-2)(s+3)} = \frac{5}{6s} + \frac{1}{10(s-2)} - \frac{14}{15(s+3)}$$

$$L^{-1}\left[\frac{5}{6s} + \frac{1}{10(s-2)} - \frac{14}{15(s+3)}\right] = \frac{5}{6} + \frac{1}{10}e^{2t} - \frac{14}{15}e^{-3t}.$$

8. $$\frac{5s+1}{s^2 - s - 12} = \frac{5s+1}{(s+3)(s-4)} = \frac{A}{(s+3)} + \frac{B}{(s-4)}.$$

Multiply out by the denominator or use cover-up rule, put $s = -3, A = 2; s = 4$, $B = 3$

$$L^{-1}\left[\frac{2}{(s+3)} + \frac{3}{(s-4)}\right] = 2e^{-3t} + 3e^{4t}.$$

9.
$$\frac{20-s}{s^2+7s+12} = \frac{20-s}{(s+3)(s+4)} = \frac{23}{(s+3)} - \frac{24}{(s+3)}$$

$$L^{-1}\left[\frac{2}{(s+3)} + \frac{3}{(s-4)}\right] = 2e^{-3t} + 3e^{4t}.$$

Also see Figure 8.2.

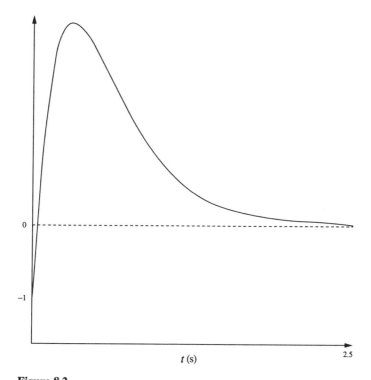

t (s)

Figure 8.2

10.

$$\frac{15}{s} \longrightarrow \boxed{\frac{10}{s+5}} \longrightarrow \frac{150}{s(s+5)}$$

$$\frac{150}{s(s+5)} = \frac{A}{s} + \frac{B}{s+5}.$$

Multiply out by the denominator or use the cover-up rule, put $s = 0, A = 30$ and $s = -5, B = -30$

$$L^{-1}\left[\frac{30}{s} - \frac{30}{s+5}\right] = 30(1 - e^{-5t}).$$

11. We can use the voltage divider rule to determine the circuit transfer function (see Figure 8.3).

$$\frac{V_o}{V_i} = \frac{\dfrac{1}{sC}}{\dfrac{1}{sC} + sL + R} = \frac{1}{s^2 LC + sRC + 1}.$$

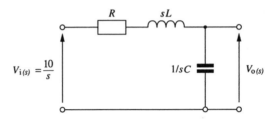

$$V_{i(s)} = \frac{10}{s} \qquad R \qquad sL \qquad 1/sC \qquad V_{o(s)}$$

Figure 8.3

We will make the coefficient of $s^2 = 1$

$$\frac{V_o}{V_i} = \frac{\dfrac{1}{sC}}{\dfrac{1}{sC} + sL + R} = \frac{1}{s^2 LC + sRC + 1}.$$

Put circuit values into the transfer function;

$$\frac{V_o}{V_i} = \frac{4000}{s^2 + 20s + 4000}.$$

V_o is the quantity required and hence we must multiply this expression by $10/s$ and divide up by partial fractions then find the inverse transforms

$$\frac{V_o}{V_i} = \frac{40{,}000}{s(s^2 + 20s + 4000)} = \frac{A}{s} + \frac{Bs + C}{s^2 + 20s + 4000}.$$

Multiply out by the denominator and equate coefficients

$$\begin{aligned} 0 &= A + B \\ 0 &= 20A + C \\ 40{,}000 &= 4000A \end{aligned}$$

therefore $A = 10$, $B = -10$ and $C = -200$

$$\frac{40,000}{s(s^2 + 20s + 4000)} = \frac{10}{s} - \frac{(10s + 200)}{s^2 + 20s + 4000}.$$

Complete the square on the denominator

$$4s^2 + 80s + 16,000$$
$$4s^2 + 80s + 16,000 + 60$$
$$4s^2 + 80s + 16,060 = (2s + 20)^2 + 15,660$$
$$= 4(s + 10)^2 + 15,660 = (s + 10)^2 + 3915.$$

This algebraic delight allows us to put

$$\frac{40,000}{s(s^2 + 20s + 4000)} = \frac{10}{s} - 10\left[\frac{(s + 10) + 10}{(s + 10)^2 + 3915}\right].$$

Now we can use the first shift theorem and replace $(s + 10)$ by s, remembering to multiply the $f_{(t)}$ by e^{-10t}, giving the following inverse Laplace transform

$$f_{(t)} = 10 - 10e^{-10t}\left[\cos\sqrt{3915}t + \frac{10}{\sqrt{3915}}\sin\sqrt{3915}t\right].$$

12.
$$\frac{2}{(s+3)(s-2)^2} = \frac{A}{s+3} + \frac{B}{(s-2)^2} + \frac{C}{s-2}.$$

Cross multiply by the denominator and equate coefficients

$$2 = A(s - 2)^2 + B(s + 3) + C(s - 2)(s + 3)$$
$$2 = As^2 - 4As + 4A + Bs + 3B + Cs^2 + Cs - 6C.$$

Equate coefficients of s:

$$(s^2) \quad 0 = A + C$$
$$(s^1) \quad 0 = -4A + B + C$$
$$(s^0) \quad 2 = 4A + 3B - 6C$$

$$A = \frac{2}{25}, \quad B = \frac{10}{25}, \quad C = -\frac{2}{25}$$

$$L^{-1}\left[\frac{2}{25(s+3)} + \frac{10}{25(s-2)^2} - \frac{2}{25(s-2)}\right] = \frac{2}{25}\left[5te^{-2t} + e^{-3t} - e^{2t}\right].$$

9 Time Response

First-order systems

All students of Control Engineering should have had an introduction to first-order systems before reaching this point in their studies. We will not dwell on this point here, you should be fairly familiar with the systems and we have had some examples of solving first-order differential equations in the previous chapter.

Remember, the first-order system will grown or decay at an exponential rate of either $A(1 - e^{-x})$ or Ae^{-x} depending on whether it represents growth or decay; whichever it is, the time taken to reach the final, or any other, value will depend on the x term which represents

$$\frac{t}{\tau},$$

where τ is the system time constant. The time constant is the time that it would take for the signal to reach its final value if it grew or decayed constantly at its original rate of change. Instead of this the energy stored in the system ensures that it only reaches 63.2% of its final value in one time constant. In the next time constant the system grows or decays 63.2% of the amount left and this continues. Mathematically the system only reaches its final value at $t = \infty$, but from a practical point of view the system will be close enough to its final value when $t = 5\tau$ (see Figure 9.1 opposite).

Second-order systems

To be able to analyse the behaviour of time-varying systems it is necessary to model these systems accurately. As engineers (and even more so as engineering students) we may not be interested in the mathematical model itself, but rather in the methods which allow us to obtain answers to questions about the system. Typical questions could be:

- How long will it take for a system to respond to a change in input?
- How accurate will the response be?
- Will the response be oscillatory, and if so how much overshoot will be present?

The nature of the output of the system will depend on the type of input; the most common input type for this purpose is the step-input, often taken as the unit-step as described in Chapter 7.

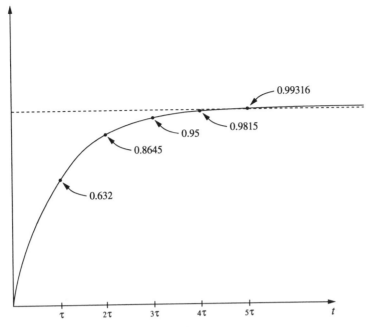

Figure 9.1 Response of the first-order system.

We can determine the response of the system to any input that it is possible to model using the Laplace transform, or we can investigate any system Transfer Function using CODAS or similar software. However, we can also use the standard Bode second-order equation to determine information from the response to a step-input. The simplicity of this approach often makes it a more appealing option.

It is useful for engineering students to have some exposure to all of these mathematical modelling methods and the connection between the parameters in a Transfer Function and the characteristic of its response to a given input signal.

Standard equation for the second-order system

The standard second-order equation can be found by mathematical analysis and is used to obtain important parameters. The standard (Bode) form for the second-order equation is

$$\frac{\omega_n^2}{s^2 + 2\zeta\omega_n + \omega_n^2}$$

and by using the equality of equations we can evaluate:

(a) The natural frequency (ω_n) of the system, which relates to the frequency (speed) of oscillation of the transient due to a change in excitation signal.
(b) The damping ratio (Greek 'zeta') which is a dimensionless representation of the amount of damping in the system.

For example, a system could be analysed by the block diagram method (or any other) and produce the following equation

$$Gp(s) = \frac{100}{s^2 + 8s + 100}.$$

When we compare this to the standard equation we can deduce:

$$\omega_n^2 = 100, \text{ hence } \omega_n = 10 \text{ rad/s}$$
$$2\zeta\omega_n = 8, \text{ hence } \zeta = 0.4.$$

But what are these parameters and what effect do they have on the system?

Systems with no damping ($\zeta = 0$)

As already mentioned, damping is the property of a body to absorb energy and the damping ratio is the ratio of the amount of damping in the system to the amount that would be required to absorb all input energy. The damping ratio or damping factor is given the symbol ζ (which does not have dimensions).

If a system has no damping whatsoever then the oscillatory nature of the response is a continuous sinusoid of frequency ω_n which neither decreases nor increases with time.

Zero damping produces 'simple harmonic motion'. Hence, the frequency ω_n is that at which the system oscillates when there is no damping present.

Underdamped systems ($0 < \zeta < 1$)

This type of system is the most common type to be investigated, or designed, by the engineer. It has a transient response which oscillates at a damped frequency of

$$\omega_d = \omega_n\sqrt{1 - \zeta^2}, \tag{9.1}$$

where ω_d is the damped natural frequency.

This difference becomes more pronounced as ζ increases in size.

Critically damped systems ($\zeta = 1$)

Here there is no oscillatory component to the transient response and the output only very slowly reaches its final value.

Overdamped systems ($\zeta > 1$)

The output of the system is now very sluggish and response time extremely slow.

Figure 9.2 on the next page shows the system's response to a step input with varying damping factors.

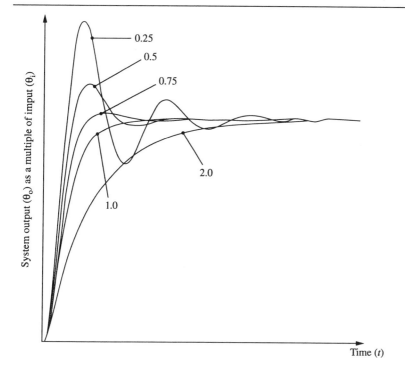

Figure 9.2

Time domain performance parameters

Dynamic performance parameters can be used to obtain:

- the degree of oscillation,
- time for the transient response to die away, and
- the response time or the time for the response to reach some prescribed fraction of the final value.

We will examine these parameters and how they can be obtained with reference to the time domain step response data of Figure 9.3.

Percentage overshoot

When we are considering oscillatory systems the amount of oscillation is normally specified by the percentage peak overshoot, or just as a percentage overshoot which is defined as

$$\text{percentage overshoot} = 100\left(\frac{P_1}{P_0}\right) \qquad (9.2)$$

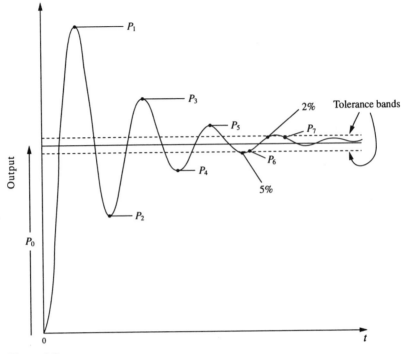

Figure 9.3

Peak time

Peak time (t_p) is the time that the response takes to reach its first peak value (which will be the only time that this value is achieved for a stable system)

$$t_p = \frac{\pi}{\omega_d} = \frac{\pi}{\omega_n \sqrt{1-\zeta^2}} \text{ seconds.} \tag{9.3}$$

The value of this maximum overshoot (P_1) can be determined

$$P_1 = e^{-\alpha},$$

where

$$\alpha = \frac{\zeta \pi}{\sqrt{1-\xi^2}}. \tag{9.4}$$

Decay ratio

The oscillations produced by an underdamped system decay within an exponential envelope and at a rate which is often defined as the ratio of two successive overshoots, normally the first two are used, i.e.

$$\text{Decay ratio} = \frac{P_3}{P_1}. \tag{9.5}$$

The magnitude of any peak can be determined from that of the first (or even from the steady-state value), by using

$$\frac{P_{k+n}}{P_k} = e^{-\alpha n}, \tag{9.6}$$

where n denotes the number of half-cycles between P_1 and P_n. Example 9.1 will demonstrate the use of this equation.

Settling time

The settling time, t_s, indicates the time it will take for the transient to die away to within some specified tolerance band. The response must not only reach, but also stay within the band. Typical tolerance bands are $\pm2\%$ and $\pm5\%$.

For the second-order system, the value of the transient component follows an envelope described by

$$1 + e^{-\zeta \omega nt} \tag{9.7}$$

which means that the curves described by these equations have time constants

$$\tau = \frac{1}{\zeta \omega_n}. \tag{9.8}$$

We can use the exponential component of this equation to determine the error – see Table 9.1.

Table 9.1

t	$e^{-\zeta \omega nt}$	Error (%)
τ	0.368	36.8%
2τ	0.135	13.5%
3τ	0.050	5.0%
4τ	0.018	1.8%
5τ	0.007	0.7%

From Table 9.1 one can see that a 5% tolerance band is reached in three time constants and a 2% band in four time constants.

To calculate the settling time to any given value we can use the exponential term

$$e^{-\zeta \omega nt} = \text{the actual value at time } (t).$$

If the tolerance band is put in as the desired value then

$$e^{-\zeta \omega nt} = 0.02 \text{ (say)}$$

$$-\zeta \omega_n t = \ln 0.02$$

$$t = \frac{\ln 0.02}{-\zeta \omega_n}.$$

If we put $\zeta = 0.3$ and $\omega_n = 5$ then

$$t = \frac{\ln 0.02}{1.5}$$

$$= 2.5 \text{ s.}$$

This can easily be checked with CODAS or some other similar software package.

Response time

Response time (t_r) is usually considered to be the time it takes for an output to go from 10% to 90% of its final value when the input is a step. Figure 9.3 shows the response time as $t_2 - t_1$. The diagram shows that a short response (or rise) time will normally be associated with a larger overshoot and hence more oscillation. In many ways this is the main problem of the system's designer or analyst, how can the minimum overshoot be optimised while achieving the response time required. As we have already seen it would take three time constants for the output to reduce to less than 5% and four time constants to reduce to less than 2%. Since the time of one cycle of oscillation is $1/f$, we can express this in terms of ω_d (which equals $2\pi f_d$)

$$\frac{2\pi}{\omega_d}.$$

In terms of the natural frequency (ω_d), where $\omega_d = \omega_n \sqrt{(1 - \zeta^2)}$ then the time for one oscillation becomes

$$t_d = \frac{1}{f} = \frac{2\pi}{\omega_n \sqrt{1-\zeta^2}}, \qquad (9.9)$$

t_d is also referred to as the period of oscillation.

We have already established that settling time to some value (x) can be determined from

$$t_s = \frac{\ln(x)}{\zeta \omega_n}. \qquad (9.10)$$

Consequently, the number of cycles required to bring the output to this arbitrary value (x) is given by

$$f_s = -\frac{\ln(x)}{\zeta\omega_n} \text{ divided by } \frac{2\pi}{\omega_d},$$

i.e. this is the settling time divided by the periodic time of the damped oscillation. This can be simplified by replacing ω_d by its identity $\omega_n\sqrt{(1-\zeta^2)}$ to give

$$f_s = -\frac{\ln(x)\sqrt{1-\zeta^2}}{2\pi\zeta}. \tag{9.11}$$

❑ Example 9.1

A second-order system is described by the following differential equation

$$1.5\frac{d^2y}{dt^2} + 20.25\frac{dy}{dt^2} + 337.5y = 337.5x.$$

The problem is to analytically determine the amplitude of the second overshoot of the response when the system is subjected to a unit step input. It is also necessary to determine the time between the application of the step and the second overshoot.

First we need to take Laplace transforms and rearrange into the standard form. Take Laplace transforms:

$$1.5s^2y_{(s)} + 20.25y_{(s)} + 337.5y_{(s)} = 337.5x_{(s)}.$$

Divide through by 1.5 to make the coefficient of $s^2 = 1$

$$s^2y_{(s)} + 13.5sy_{(s)} + 225y_{(s)} = 225x_{(s)}$$

$$y_{(s)}(s^2 + 13.5s + 225) = 225x_{(s)}$$

$$\frac{y}{x}(s) = \frac{225}{s^2 + 13.5s + 225}.$$

Now compare with the standard form equation

$$\omega_n = \sqrt{225} = 15 \text{ rad/s}$$
$$2\zeta\omega_n = 13.5, \text{ therefore } \zeta = 0.45.$$

The first overshoot $P_1 = e^{-\beta}$ [eqn (9.6)] which is equal to 0.205346 over the unit step and hence the actual magnitude is 1.205346.

The second overshoot is P_3 and hence $n = 2$ in eqn (9.6): $P_3 = 0.009$ above the unit value.

We have seen that we can use the relationship $e^{-\zeta\omega n}$ = magnitude of output above or below the final value, so now put in values:

$$e^{-6.75t} = 0.009$$

$$-6.75t = \ln(0.009)$$

$$t = \frac{\ln(0.009)}{-6.75}$$

$$t = 697.8 \text{ ms.}$$

See Figure 9.4.

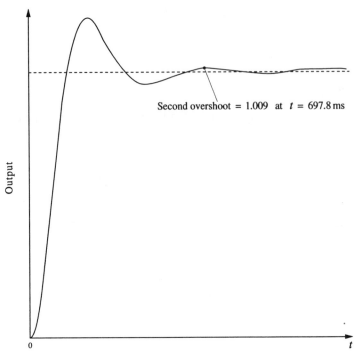

Second overshoot = 1.009 at t = 697.8 ms

Figure 9.4

System response

We have already seen that we can deduce parameters from the second-order equation and that these parameters can be used to describe the behaviour of a system under certain conditions. Laplace transforms can also be used to determine an expression for the magnitude of the system output at any value of time. In order to do this we need to multiply the standard form equation by the transform of a unit step (or any other value of step) and then use partial fractions and inverse transforms to determine the time-domain output

$$\frac{1}{s} \cdot \frac{\omega_n^2}{s^2 + 2\zeta\omega_n s + \omega_n^2} = \frac{\omega_n^2}{s(s^2 + 2\zeta\omega_n + \omega_n^2)}$$

$$\frac{\omega_n^2}{s(s^2 + 2\zeta\omega_n s + \omega_n^2)} = \frac{A}{s} \cdot \frac{Bs + C}{s^2 + 2\zeta\omega_n s + \omega_n^2}.$$

Multiply both sides by the denominator

$$\omega_n^2 = A(s^2 + 2\zeta\omega_n s + \omega_n^2) + (Bs + C)s$$
$$\omega_n^2 = As^2 + 2A\zeta\omega_n s + A\omega_n^2 + Bs^2 + Cs.$$

Equate coefficients of s

$$(s^2) \quad 0 = A + B$$
$$(s^1) \quad 0 = A(2\zeta\omega_n) + C$$
$$(s^0) \quad \omega_n^2 = A\omega_n^2$$

which gives

$$A = 1, \ B = -1, \ C = -2\zeta\omega_n.$$

Hence

$$\frac{\omega_n^2}{s(s^2 + 2\zeta\omega_n s + \omega_n^2)} = \frac{1}{s} - \frac{s + 2\zeta\omega_n}{s^2 + 2\zeta\omega_n s + \omega_n^2}$$
$$= \frac{1}{s} \cdot \frac{(s + \zeta\omega_n) + \zeta\omega_n}{(s + \zeta\omega_n)^2 + \omega_d^2}.$$

Use the shift theorem to replace $(s + \zeta\omega_n)$ with s and multiply the inverse transform by $e^{-\zeta\omega nt}$ to give an inverse transform

$$\theta_o = \theta_i \left[1 - \frac{e^{-\zeta\omega nt}}{\beta} \cdot \left\{ \sin(\omega_d t + \phi) \right\} \right],$$

where

$$\beta = \sqrt{(1 - \zeta^2)}$$
$$\omega_d = \omega_n \beta$$
$$\theta = \cos^{-1} \zeta.$$

❑ Example 9.2

Consider a Control System represented by the expression

$$\frac{3}{s^2 + \sqrt{3}s + 3}$$

and determine the output 4.2 s after the application of a unit step

$$\omega_n = \sqrt{3}, \ \zeta = 0.5.$$

Remember that θ is in radians

$$\theta_o = 0.9734 \text{ at } t = 4.2 \text{ s.}$$

See Figure 9.5.

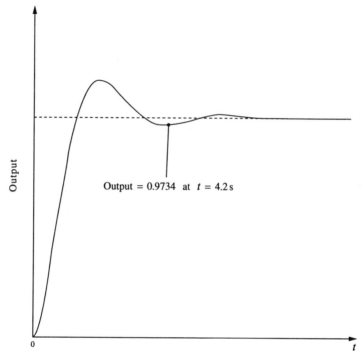

Output = 0.9734 at $t = 4.2\,\text{s}$

Figure 9.5

❏ *Example 9.2*

1. A second-order system is described by the following differential equation

$$1.5\frac{d^2y}{dt^2} + 1.2\frac{dy}{dt} + 6y = 6x.$$

Determine analytically the maximum percentage overshoot of the response and the time taken to reach this overshoot when the system input (*y*) is a unit step.

2. (a) With the aid of diagrams explain the meaning of the following terms when applied to the response of a second-order control system:
   (i)   peak overshoot,
   (ii) time to first peak,
   (iii) settling time.
   (b) A second-order position control system has a damping ratio of 0.5 and an undamped natural frequency of 6 rad/s. Determine, following the application of a unit step input:
   (i)   an expression for the output as a function of time,
   (ii) the value of the percentage peak overshoot,
   (iii) the value of the output at $t = 1\,s$.

3. Find both the undamped natural frequency and the damped frequency, the damping ratio, and the predominant time constant for a system described by the equation

$$\frac{d^2\theta_o}{dt^2} + 5\frac{d\theta_o}{dt} + 7\theta_o = 7\theta_i.$$

4. The transfer function of a second-order system is given by

$$G_{(s)} = \frac{3}{s^2 + 0.4s + 3}.$$

Analytically determine the amplitude of the third overshoot of the response to a unit step and the time between the application of this input and the third overshoot.

5. Show that the 10% to 90% rise time for a first-order system is about 2.2 time constants.

Further investigation of the second-order RLC circuit

The RLC electrical circuit (Figure 9.6) is a very useful model to demonstrate these methods since it can be easily investigated in any college laboratory; a description of such an investigation is included in the appendix. First, let us explore the circuit and see what we can expect to find.

We can replace $j\omega$ with s and operate in the s-plane. Let us obtain an expression for the circuit current when the switch is closed.

$$I_{(s)} = \frac{E}{s} \cdot \frac{1}{R + sL + \dfrac{1}{sC}}.$$

Divide through by L and rearrange

$$I(s) = \frac{E}{L} \cdot \frac{1}{s^2 + \dfrac{(R/L)s + 1}{LC}}.$$

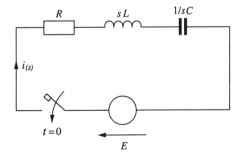

Figure 9.6

This can be fitted to the Laplace transform pair

$$F(s) = A \cdot \frac{1}{s^2 + 2as + \omega_n^2}$$

which transforms from

$$f(t) = \frac{A}{\omega_d} \cdot e^{-at} \sin \omega_d \cdot t,$$

where wn = $\sqrt{(\omega_d^2 + \alpha^2)}$. Hence, it can be concluded that

$$\frac{E}{L} = A$$

$$\frac{R}{L} = 2a \text{ and } a = \frac{R}{2L}$$

$$\frac{1}{LC} = \omega_n^2 \text{ and } \omega_n = \frac{1}{\sqrt{LC}}$$

which means for our RLC circuit that

$$i(t) = \frac{E}{\omega_d \cdot L} \cdot e^{-at} \sin \omega_d \cdot t.$$

If this were to be displayed pictorially it would show the current as an exponentially decaying sine wave, describing an oscillatory system with a damped frequency of response ω_d. The symbol a is often known as the damping coefficient.

Appendix

□ *Example 9.3*

1. For a second-order equation described by

$$1.5\frac{d^2y}{dt^2} + 1.2\frac{dy}{dt} + 6y = 6x,$$

take Laplace transforms (the initial conditions are zero)

$$1.5s^2y_{(s)} + 1.2sy_{(s)} + 6y_{(s)} = 6x_{(s)}$$

divide through by 1.5 to make the coefficient of $s_2 = 1$ and rearrange

$$\frac{y}{x}(s) = \frac{4}{s^2 + 0.8s + 4}.$$

Compare this to standard second-order equation and deduce

$$\omega_n = \sqrt{4} = 2 \text{ rad/s}$$
$$2\zeta\omega_n = 0.8 \text{ therefore } \zeta = 0.2.$$

First peak $= P_1 = e^{-\alpha}$ [eqn (9.4)] $= 0.5266$ error $= 1.5266$. From eqn (9.10)

$$t = \frac{\ln(0.5266)}{-0.4}$$
$$= 1.6032 \text{ s (see Figure 9.7)}.$$

2. (a) A solution to this part of the example can be taken directly from the text.
   (b) (i) As $\omega_n = 6$ rad/s and $\zeta = 0.5$ then we can use the following equation

$$\theta_o = \theta_i\left[1 - \frac{2e^{-3t}}{\sqrt{3}} \cdot \sin\left[3\left(\sqrt{3}\right)t + \frac{\pi}{3}\right]\right].$$

   (ii) $P_1 = e^{-\alpha}$ [(eqn 9.4)] $= 0.163 = 16.3\%$.
   (iii) By calculation: $\theta_o = 1.0022895$ at $t = 1$ s.

3. Take Laplace transforms:

$$s^2\theta_{o(s)} + 5s\theta_{o(s)} + 7\theta_{o(s)} = 7\theta_{i(s)}$$

$$\frac{\theta_o}{\theta_i}(s) = \frac{7}{s^2 + 5s + 7}.$$

Hence we can deduce that

$$\omega_n = \sqrt{7} \text{ rad/s (undamped natural frequency)}$$

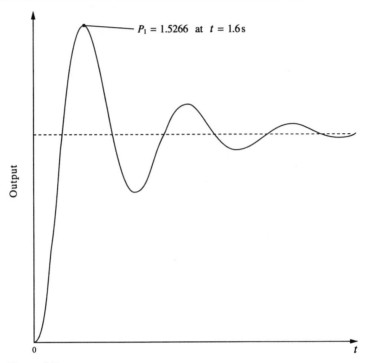

$P_1 = 1.5266$ at $t = 1.6$ s

Output

0

t

Figure 9.7

$$\zeta = 0.945 \text{ (damping ratio)}$$
$$\omega_d = 0.866 \text{ rad/s (damped frequency)}$$
$$\tau = 0.84664 \text{ s (predominant time constant).}$$

4.
$$Gp(s) = \frac{3}{s^2 + 0.4s + 3}.$$

Hence, $\omega_n = \sqrt{3}$ and $2\zeta\omega_n = 0.4$, therefore $\zeta = 0.11547$. The first overshoot $= P_1$ $= e^{-0.3652} = 0.694$ [from eqn (9.4)]. We can then use this as we want a value for P_5 and hence $n = 4$ in eqn (9.6)

$$P_5 = 0.161$$

and

$$ts = \frac{\ln(0.161)}{-\zeta\omega_n} = 9.1305 \text{ s.}$$

5. See Figure 9.8.

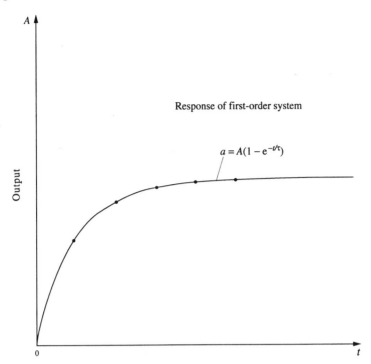

Figure 9.8

$$a = A(1 - e^{-t/\tau})$$

$$\frac{a}{A} = 1 - e^{-t/\tau}$$

$$e^{-t/\tau} = 1 - \frac{a}{A}$$

$$-\frac{t}{\tau} = \ln\left(1 - \frac{a}{A}\right)$$

$$t = -\tau \ln\left(1 - \frac{a}{A}\right).$$

At t_1, $a = 10\%A$ and at t_2, $a = 90\%A$. Put these values into the equation and then t_1 (approximately) $= 0.1054\tau$ and t_2 (approximately) $= 2.303\tau$. Subtract these and $t_2 - t_1 = $ (approx) 2.2τ.

Practical task: under-damped system parameters

You will need the following equipment:

- oscilloscope
- signal generator
- proto-board
- 10 kΩ variable resistor, R (a decade box is ideal)
- 100 mH inductor, L
- 33 nF capacitor, C

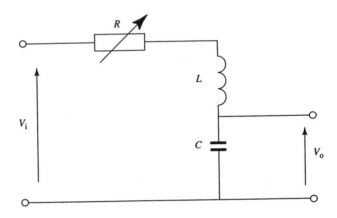

Method

1. Construct the circuit as shown in the diagram.
2. Set the signal generator to 'square' at 200 Hz, 2 V peak and set R to 1 kΩ. If the resistance of the inductor is negligible then this value will equal the total circuit resistance r.
3. Display and record the output waveform across the capacitor, C, for various values of R.
4. For the various values of R, measure (or estimate) and tabulate values for:

ω_d [eqn (9.1)] and t_d [eqn (9.9)]
percentage overshoot [eqn (9.2)]
peak time t_p [eqn (9.3)]
settling time t_s [eqn (9.10)]
response time t_r

5. Compare practical parameters to those predicted from theory:

$$\xi = \frac{r}{2L}; \quad \omega_n = \frac{1}{\sqrt{(LC)}} = \text{eqn (9.1)}.$$

10 Pole Zero analysis

We have seen that systems can be modelled by Transfer Functions which are very often in the form of a ratio between two polynomials. The polynomials are in terms of the s operator and this enables us to map the roots of the equation onto the s-plane. Remember that

$$s = \sigma + j\omega,$$

where σ and ω are real variables. If a function of s [$F(s)$] describes a system, and if

$$F(s) = \frac{(s - Z_1)(s - Z_2) \dots}{(s - P_1)(s - P_2) \dots}.$$

any value of Z which makes the numerator equal to zero is called a **zero** and any value of P which makes the denominator equal to zero is called **pole**. These poles and zeros can be mapped onto the s-plane and as ω varies their position will vary.
This can be illustrated by the following example.

❑ *Example 10.1*

Determine the poles and zeros of a Control System described by the following Transfer Function

$$G(s) = \frac{s^2 + 7s + 10}{s^3 + 4.5s^2 + 13.5s + 35}.$$

We want the roots of both the numerator and the denominator. The numerator is a quadratic and so we can factorise it or use the quadratic equation to find the roots.

$$\frac{-7}{2} \pm \frac{\sqrt{7^2 - 4(1)(10)}}{2}$$

$$-3.5 \pm 1.5$$

$$= -5 \text{ or } -2$$

remember that these are the values of s which will make the equation go to zero, i.e. $s^2 + 7s + 10 = (s + 5)(s + 2)$. Zeros exist at $s = -5$ and $s = -2$; the roots of the denominator now need to be found.
The following equation

$$s^3 + 4.5s^2 + 13.5s + 35$$

is a cubic equation with all real coefficients and hence we can be sure that the equation has at least one real root. Also, since all the coefficients are positive then the real root must be negative and to find that root would leave a quadratic equation. There are several method available to find the root: Newton's method will be used here.

The locality of the real root can be determined by putting various negative values to s and observing when a change of sign happens.

$$
\begin{aligned}
s = 0: \quad & f(s) = 35 \\
s = -1: \quad & f(s) = -1 + 4.5 - 13.5 + 35 && = 25 \\
s = -2: \quad & f(s) = -8 + 18 - 27 + 35 && = 18 \\
s = -3: \quad & f(s) = -27 + 40.5 - 40.5 + 35 && = 8 \\
s = -4: \quad & f(s) = -64 + 72 - 54 + 35 && = -11.
\end{aligned}
$$

We see that the sign has changed and hence a real root (perhaps not always the only one) lies between -3 and -4. Newton's method states that, given a first estimate of the root s^1 then a second estimate s^2 is obtained by

$$s^2 = s^1 - \frac{f(s^1)}{f'(s^1)},$$

where $f'(s^1)$ is the first differential of $f(s)$ with $s = s^1$.
Let us take a first estimate for the root at $s = -3.4$

$$f(-3.4) = (-3.4)^3 + 4.5(-3.4)^2 + 13.5(-3.4) + 35 = 1.816$$
$$f'(s) = 3s_2 + 9s + 13.5$$
$$f'(-3.4) = 3(-3.4) + 9(-3.4) + 35 = 16.68$$

$$s^2 = -3.4 - \frac{1.816}{16.68} = -3.5.$$

Let us try this as the second estimate s_2

$$f(-3.5) = (-3.5)^3 + 4.5(-3.5)^2 + 13.5(-3.5) + 35 = 0$$

Hence, if the denominator polynomial is divided by $(s + 3.5)$ a quadratic will be obtained which we can factorise by the equation, or otherwise, giving

$$G(s) = \frac{(s+2)(s+5)}{\left(s+3.5\right)\left(s+\frac{1}{2}+j\sqrt{[39/4]}\,\right)\left(s+\frac{1}{2}-j[39/4]\right)}.$$

Consequently, it can be seen that this function has **zeros** at -2 and -5 and **poles** at -3.5, $-1/2 + j\sqrt{[39/4]}$ and $-1/2 - j\sqrt{[39/4]}$ and we can plot these onto the s-plane.

The s operator $= \sigma + j\omega$ and so σ is real whilst $j\omega$ will map onto the imaginary axis (see Figure 10.1).

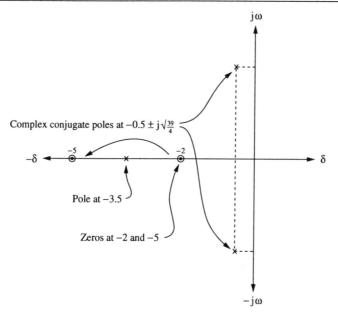

Figure 10.1

Now that we have some idea of how to determine poles and zeros let us examine what can be deduced from them.

Stability

The characteristic equation of a system can be obtained by putting the denominator of the **closed-loop** transfer function equal to zero. Once obtained, the roots of this equation must be negative for the system to be stable. This means that the poles must appear in the left-hand half of the s-plane.

The system gain constant K will affect the value of the closed-loop poles and a graphical root-locus technique illustrates the variation in the position of the closed-loop poles with the variation of K. This technique allows the designer to use an optimum value of K.

Remember that the open-loop transfer function is of the form

$$G(s) = \frac{K}{P(s)},$$

where P is a polynomial in the s-plane. The closed-loop transfer function is

$$G(s) = \frac{K}{P(s) + K}$$

and the characteristic equation is

$$P(s) + K = 0$$

which will have roots depending upon the value of K.

Let us illustrate this with the following example.

❏ *Example 10.1*

A system has an open-loop transfer function

$$G(s) = \frac{K}{s(s+2)}$$

which means that the closed-loop transfer function is

$$G(s) = \frac{K}{s(s+2) + K}$$

and the characteristic equation is: $s(s + 2) + K = 0$ with poles at positions determined by K. When $K = 1$, there are two poles at $s = -1$, i.e.

$$s^2 + 2s + 1 = 0$$
$$(s + 1)(s + 1) = 0$$

when $K = 2$ there are conjugate complex poles at $s = -1 \pm j$; when $K = 4$ there are complex conjugate poles at $s = -1 \pm j\sqrt{3}$; when $K = 5$, $s = -1 \pm j2$; and so on. These variations can be plotted onto the s-plane and will produce a root-locus.

There are a number of rules which can be used to make the plotting of the root-locus easier. However, software such as CODAS will plot this for us in an automatic manner and again we see the advantage of computer methods. However, it is still important to go through a 'hand calculation' example in order to obtain an understanding of the method.

Root-locus rules

1. Each locus starts at an open-loop pole when $K = 0$ and either ends at an open-loop zero or it approaches infinity when $K = \infty$.
2. There will be the same number of branches of the loci as there are open-loop poles.
3. The loci will always be symmetrical about the real axis (σ-axis).
4. If the open-loop transfer function has m open-loop poles and n open-loop zeros and if $n > m$ then we can say that there will be $m - n$ branches of the loci approach infinity.
5. If there are an odd number of poles and/or zeros then the real axis to the left of these will form part of the loci.
6. We will want to know where the asymptotes cross the real axis and we can determine this point x by using the following equation

$$x = \frac{(\Sigma \text{open-loop poles} - \Sigma \text{open-loop zeros})}{(m - n)}.$$

7. The angle which the asymptotes make with the real axis can be calculated using

$$\alpha = \pm \frac{k(180)}{m-n},$$

where k is an odd integer.

8. The value of K at the crossing point of the $j\omega$ axis can be determined by the Routh criterion or from rule 9.

9. As in rule 8, these crossing points can be found by
   (i) replace s by $j\omega$ in the characteristic equation,
   (ii) put the j terms equal to zero,
   (iii) solve for both ω and K.

10. There will always be a breakaway point between any two adjacent poles on the real axis which are connected together by a branch of the loci.

11. The loci breaks away from the real axis where K is a maxima. This maximum can be found from the maxima of the characteristic equation.

12. The angle at which the loci leaves the complex poles can be determined by the relation $\Sigma\theta_z - \Sigma\theta_p = \pm 180°$. Here, $\Sigma\theta_z$ is the sum of the angles subtended by the zeros and $\Sigma\theta_p$ is the sum of the angles subtended by the zeros.

❑ *Example 10.2*

Consider a unity feedback control system with the following open-loop transfer function

$$G(s) = \frac{K}{s(s+1)(s+2)}.$$

Sketch the root locus and while noting the connection between ζ and K, determine a value of K such that $\zeta = 0.5$. We will apply the rules as listed and compare the sketched root-locus with that produced by CODAS. This system has open-loop poles and values of s which will drive the gain to infinity, at $s = 0, -1$ and -2. The characteristic equation is the denominator of the closed-loop transfer function (with unity feedback), equal to zero. In this case we can obtain the characteristic equation by putting

$$G = \frac{K}{s(s+1)(s+2)}$$

and then finding the closed-loop transfer function

$$\frac{G}{1+G}$$

and equating its denominator to zero

$$s_3 + 3s^2 + 2s + K = 0.$$

In this case all three poles are real and hence part of the negative real axis will form part of the loci (rule 5).

The angles that the asymptotes depart from the real axis can be determined from rule 7. In our case $m = 3$ and $n = 0$ (rule 4). Therefore, the angle of departure is

$$\beta = \pm \frac{k_{180}}{m-n}, \quad k = 1, \ 3$$

when $k = 1$, $\beta = \pm 60°$ and when $k = 3$, $\beta = \pm 180°$. The intersection with the real axis can be determined from rule 6

$$x = \frac{(\Sigma \, \text{open-loop} \ \text{poles} - \Sigma \, \text{closed-loop} \ \text{poles})}{m-n}$$

$$x = \frac{0-1-2}{3} = -1.$$

The points at which the loci cross the $j\omega$ axis can be determined from the Routh array (details of which are given in the appendix)

$$\begin{bmatrix} s^3 & 1 & 2 \\ s^2 & 3 & K \\ s^1 & \dfrac{6-K}{3} & \\ s^0 & K & \end{bmatrix}.$$

Hence, the limiting value of K for stability is 6 and so the intersection with the imaginary axis is found from

$$3s^2 + 6 = 0$$

or

$$s = \pm j\sqrt{2}.$$

We can use this information to plot the asymptotes and compare these to the computer graphic as shown in Figures 10.2 and 10.3.

For any given value of K the complex conjugate poles can be plotted as shown in Figure 10.4 on page 110. As K varies so the angle Φ will also vary. The cosine of this angle is the damping ratio and so for a damping ratio of $\zeta = 0.5$, the angle $\Phi = 60°$ and gain constant $K = 1.037$.

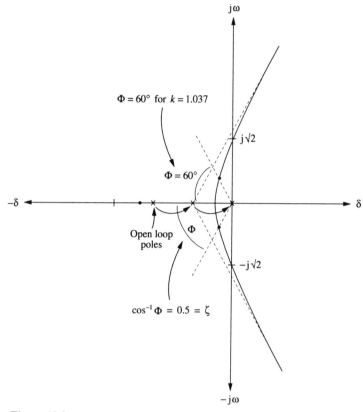

Figure 10.2

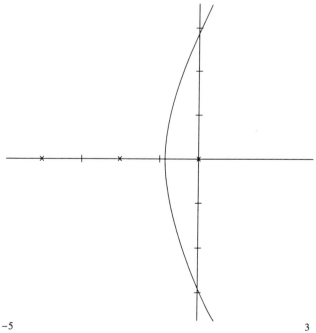

Figure 10.3 Computer plot of root locus for Example 10.2.

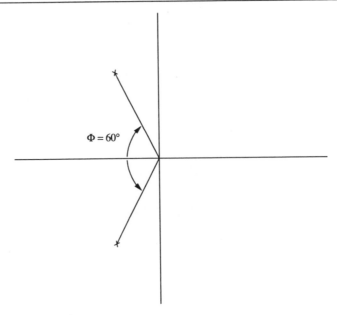

Figure 10.4

Determination of transient response

Once the root-locus has been drawn it can be used to select a value of system gain K such that the transient behaviour in the closed-loop state is desirable. The normal criteria are stability and the percentage overshoot. As the gain is increased, the dominant complex poles move further from the negative real axis and closer to the imaginary axis. If this migration is unabated, the system will become highly oscillatory and, as the pole crosses into the right-half of the s-plane, eventually unstable. The designer would wish to choose K such that the system exhibits reasonable transient behaviour. This is achieved when the angle subtended by the dominant poles to the negative real axis is between 60° and 30°.

The damping ratio of the system (ζ) is the cosine of the angle subtended by the dominant poles to the negative real axis, as illustrated in the previous example. We can see that the transient behaviour is determined by the position of the s-plane poles.

❏ *Example 10.3*

Sketch the root locus of a system with open-loop transfer function

$$G(s)\frac{K}{s(s^2 +6s+8)}.$$

From the sketch estimate the maximum value of system gain factor for stability, compare your sketch and solution with that produced by CODAS.

Appendix: Pole Zero analysis

The Routh–Hurwitz stability criterion

The Routh–Hurwitz criterion gives us a method by which we can determine the roots of the s polynomial and if they are in the right- or left-hand half of the s-plane, without actually finding the roots.

There are two necessary conditions which must be satisfied:

1. All coefficients in the polynomial have the same sign.
2. None of the coefficients are zero.

Consider the following polynomial

$$P(s) = sn + a_1 s^{n-1} + a_2 s^{n-2} + \ldots + a_n.$$

The Routh array contains $(n + 1)$ rows. The first two of these contain the coefficient of the polynomial in the following manner

$$\begin{bmatrix} s^n & 1 & a_2 & a_4 & a_6 \\ s^{n-1} & a_1 & a_3 & a_5 & a_7 \end{bmatrix}.$$

Then

$$\begin{bmatrix} b_1 & b_2 & b_3 & \cdots & \cdots \\ c_1 & c_2 & \cdots & \cdots & \cdots \\ d_1 & d_2 & \cdots & \cdots & \cdots \\ e_1 & \cdots & \cdots & \cdots & \cdots \end{bmatrix},$$

where

$$b_1 = \frac{a_1 \times a_2 - a_1 \times a_3}{a_1}$$

$$b_2 = \frac{a_1 \times a_4 - a_1 \times a_5}{a_1}$$

$$c_1 = \frac{b_1 \times a_3 - a_1 \times b_2}{b_1}$$

$$c_2 = \frac{b_1 \times a_5 - a_2 \times b_3}{b_1}, \text{ etc.}$$

We will use the polynomial of the Example 10.2

$$a_1 = 3, \ a_2 = 2, \ a_3 = K$$

$$b_1 = \frac{6-K}{3}$$

$$b_2 = 0, \ c_1 = K, \ c_2 = 0$$

$$
\begin{bmatrix}
s^3 & 1 & 2 & 0 & 0 \\
s^2 & 3 & K & 0 & 0 \\
s^1 & \dfrac{6-K}{3} & 0 & 0 & 0 \\
s^0 & K & 0 & 0 & 0
\end{bmatrix}.
$$

From this array we can deduce that the limiting value of K for stability is 6 since for any value larger than this b_1 becomes negative and that would mean instability. The points of intersection with the imaginary axis can be found from

$$3s^2 + 6 = 0$$

therefore

$$s = \pm j\sqrt{2}.$$

Solution to examples

The characteristic equation of this example is

$$s^3 + 6s^2 + 8s + K = 0.$$

Applying the rules previously discussed

1. Open-loop poles at $s = 0$, $s = -4$ and $s = -2$.
2. No zeros and hence all loci move to infinity.
3. There are three open-loop poles and hence three loci branches (rule 2).
4. An odd number of poles means part of loci will lay on the negative real axis (rule 5).
5. Point of real-axis departure

$$= \frac{0-2-4}{3} = -2 \ \text{(rule 6)}.$$

6. Angle of departure from the real axis

$$= \pm \frac{1(180)}{3} = 60° \ \text{(rule 7)}$$

7. The Routh array is as follows:

$$\begin{bmatrix} s^3 & 1 & 8 & 0 \\ s^2 & 6 & K & 0 \\ s^1 & \dfrac{49-K}{6} & 0 & 0 \\ s^0 & K & 0 & 0 \end{bmatrix}.$$

From this we can deduce that no negative terms in first column means that there are no poles in the right half of the s-plane, hence stable behaviour exists. This fixes the maximum value of K for stability since should it exceed 48 then the previous statement is not true.

8. Points of intersection with the imaginary axis are found from $6s^2 + 48 = 0$. Therefore $s = \pm j2.82$.

We can use this information to sketch the root locus as shown in Figure 10.5.

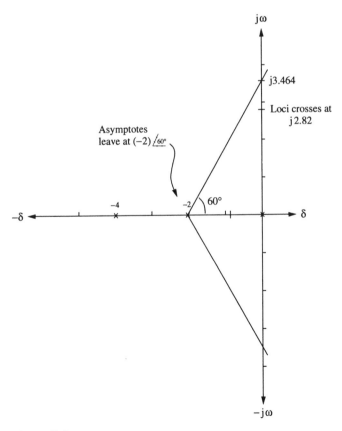

Figure 10.5

11 Worked examples

As a final chapter a few worked examples are given. These examples are not in any particular order and are random in their degree of difficulty. The solutions are separate to the problems so that readers, if they wish, can work through them first only checking the solution if and when required.

1. For the system represented by the block diagram in Figure 11.1 determine
   (i) the open and closed-loop transfer functions
   (ii) and the time domain response to a unit-step input.
2. A potentiometer is connected to a d.c. supply of 100 V and is capable of total usable rotation of 1.4π radians. What is the effective Transfer Function of the potentiometer?
3. A Control System can be described by the following second-order differential equation

$$\frac{d^2\theta_0}{dt^2} + 2\zeta\omega_n\frac{d\theta_0}{dt} + \omega_n^2\theta_0 = k\omega_n^2\theta_i.$$

Given that the undamped natural angular velocity of the system is 10 rad/s and that the system has a damping ratio of 0.6 determine the gain and phase of the system when an angular velocity of 20 rad/s is applied to the system ($k = 5$).

4. The block diagram shown in Figure 11.2 has the following block transfer functions

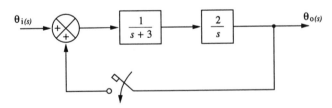

Figure 11.1

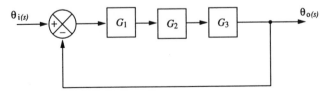

Figure 11.2

$$G_1 = \frac{2}{1+s\tau_1}, \quad \tau_1 = 50 \text{ ms}$$

$$G_2 = 15$$

$$G_3 = \frac{5}{1+s\tau_2}, \quad \tau_2 = 200 \text{ ms.}$$

Determine the closed-loop Transfer Function of the system.

5. If a first-order system has the following function

$$\frac{V_0}{V_i}(s) = \frac{100}{s+20}$$

determine the output voltage 80 ms after the application of a 5 V step input.

6. How long will it take the output of a first order system to reach three-quarters of its final output value if its time constant is 14 s?

7. If a second-order system has the following transfer function

$$G(s) = \frac{100}{s^2 + 5s + 100}$$

determine the magnitude of the third overshoot of the oscillating output following the application of a unit step. Further, calculate the time it would take for the system to settle to within 2% of its final value.

8. Sketch a Bode plot for the system with the following Transfer Function

$$G(s) = \frac{30}{s(0.5s+1)}$$

and estimate the gain and phase margins of the system.

9. With reference to Figure 11.3:
   (i) solve for y in terms of x, d and H,
   (ii) solve for y if $x = 0.2$, $d = 0.02$ and $H = 0.1$.

10. What is the output of the network shown in Figure 11.4 when

$$\omega = \frac{1}{\tau} \text{rad s}^{-1}$$

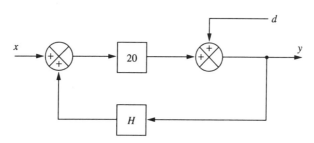

Figure 11.3

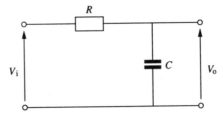

Figure 11.4

11. A telescope antenna positional control system is modelled by the following differential equation

$$\frac{d^2\theta_o}{dt^2} + 4\frac{d\theta_o}{dt} + 100\theta_o = 100\theta_i$$

(i) determine the undamped natural angular velocity of the system and its damping factor,

(ii) what value of angular position will the system have achieved at $t = 0.5$ s after the application of a sudden input of $\pi/3$ radians.

12. Sketch Bode plots for the following functions:

(a) $\dfrac{1 + s\tau_1}{1 + s\tau_2}$

(b) $\dfrac{1}{1 + s\tau_3}$

(c) $1 + s\tau_2$

(d) $20/s$,

where

$$\tau_1 = 0.1 \text{ s}, \quad \tau_2 = 0.5 \text{ s}, \quad \text{and} \quad \tau_3 = 2.0 \text{ s}.$$

13. Write down an s domain Transfer Function for the circuit shown in Figure 11.5 and then solve for both gain and phase of the function when the angular velocity is 1.5 rad/s.

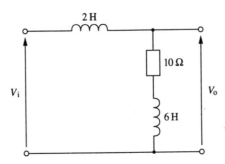

Figure 11.5

14. (a) Write down Laplace transforms for the following functions:
    (i) 3 sin 5t,
    (ii) $1 - e^{-t/\tau}$,
    (iii) $1/3\ e^{-2t}$,
    (iv) 10 sin 6t + 20 cos 5t.
  (b) Find the functions of time represented by the functions of s below

   (i) $\dfrac{3+2s}{s(s+3)}$

   (ii) $\dfrac{A}{s(s\tau+1)}$, $(A = 4,\ \tau = 0.6)$

   (iii) $\dfrac{2}{3s+2}$

   (iv) $\dfrac{4}{s^2 +6s+100}$.

15. Given that

$$\frac{\theta_o}{\theta_i} = \frac{10}{2s+3}$$

   find the system gain and phase at 6 rad/s.

16. For the circuit shown in Figure 11.6 determine the value of the circuit current 50 ms after the switch is closed.

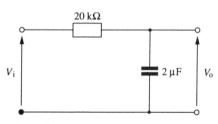

Figure 11.6

17. If a rudder control mechanism moves $+ (\pi/9)$ rad following the application of a 5 V command signal, what function can be used to describe the controller?

18. Considering Figure 11.7

$$\tau_a = \frac{1}{1+0.05s},\quad \tau_b = \frac{1}{1+0.2s},\quad A = 100$$

   (i) determine the gain and phase of the open-loop system when $\omega = 50$ rad/s,
   (ii) if the system is now closed-loop with unity feedback determine its gain and phase at the angular velocity of part (i).

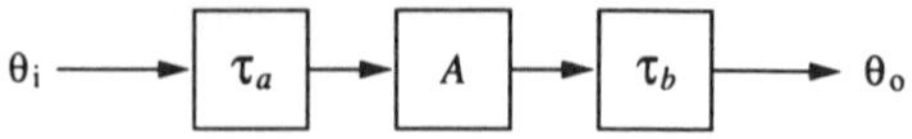

$$\theta_i \longrightarrow \boxed{\tau_a} \rightarrow \boxed{A} \rightarrow \boxed{\tau_b} \longrightarrow \theta_o$$

Figure 11.7

Solutions to the examples

1. (i) Open loop

$$= \frac{2}{s(s+3)}.$$

Closed loop

$$\frac{\dfrac{2}{s(s+3)}}{1 - \dfrac{2(4s)}{s(s+3)}}$$

$$= \frac{2}{s(s-3)-8s}$$

$$= \frac{2}{s(s-5)}.$$

(ii) $\dfrac{1}{s} \cdot \dfrac{2}{s(s-5)} = \dfrac{2}{s^2(s-5)} = \dfrac{A}{s^2} + \dfrac{B}{s} + \dfrac{C}{s-5}$

multiply both sides by $s^2(s-5)$

$$2 = A(s-5) + Bs(s-5) + Cs^2$$
$$2 = As - 5A + Bs^2 - 5Bs + Cs^2.$$

Equate coefficients of s

$$[S^2] \quad 0 = B + C$$
$$[S^1] \quad 0 = A - 5B$$
$$[S^0] \quad 2 = -5A \text{ therefore}$$

$$A = -\frac{2}{5}, \ B = -\frac{2}{25}, \ C = \frac{2}{25}.$$

Hence

$$\frac{2}{s^2(s-5)} = \frac{2}{25(s-5)} - \frac{2}{25s} - \frac{2}{5s^2}$$

$$L^{-1} = \frac{2}{25}e^{st} - \frac{2}{2s} - \frac{2}{5}t.$$

2. $\theta_i = 100$ V and $\theta_o = 1.4\pi$ rad, therefore

$$\frac{\theta_o}{\theta_i} = \frac{1.4\pi}{100}$$

3. Take Laplace transforms and rearrange for the transfer function

$$s^2\theta_{o(s)} + 2\zeta\omega_n s\,\theta_{o(s)} + \omega_n^2\,\theta_{o(s)} = k\omega_n^2\,\theta_{i(s)}$$

$$\theta_{o(s)}(s^2 + 2\zeta\omega_n s + \omega_n^2) = k\omega_n^2\,\theta_{i(s)}$$

$$\frac{\theta_o}{\theta_i}(s) = \frac{k\omega_n^2}{s^2 + 2\zeta\omega_n s + \omega_n^2} = \frac{100k}{s^2 + 12s + 100}$$

put $s = j\omega$, and $\omega_n = 20$ with $k = 5$

$$\frac{\theta_o}{\theta_i}(j\omega) = \frac{500}{(j20)^2 + j240 + 100} = \frac{500}{-300 + j240}$$

$$= 1.3\angle -141.34°.$$

4.
$$G(s) = \frac{150}{(1+0.05s)(1+0.2s)} \quad \text{(feedforward transfer function)}$$

as the negative feedback is unity, therefore

$$G(s) = \frac{G}{1+G} = \frac{150}{0.01s^2 + 0.25s + 151}$$

can be written as

$$= \frac{15,000}{s^2 + 25s + 15,100}.$$

5. $V_{o(s)} = V_{i(s)} \cdot \text{TF}$

$$\frac{5}{s} = \frac{100}{s+20} = \frac{500}{s(s+20)}$$

see Laplace transform pair number 6.
    $\alpha = 20$ and to find the multiplying factor k

$$k20 = 500 \therefore k = 25$$

$$L^{-1}\left[25\left[\frac{20}{s(s+20)}\right]\right] = 25(1-e^{-20t})$$

put $t = 80$ ms and $V_o = 19.953$ V.

6. A first-order system will have an output given by

$$\theta_o = \theta_i (1 - e^{-t/\tau})$$

and here $\theta_o = 0.75\theta_i$ hence $0.75\theta_i = \theta i (1 - e^{-t/\tau})$

$$e^{-t/\tau} = 1 - 0.75$$
$$-t/\tau = \ln (0.25) \text{ (remember } \tau = 14 \text{ s)}$$
$$t = 19.4 \text{ s.}$$

7. $$\frac{100}{s^2 + 5s + 100}$$

compare to

$$\frac{k\omega_n^2}{s^2 + 2\zeta\omega_n s + \omega_n^2}$$

$\omega_n = 10$ rad s^{-1} and $\zeta = 0.25$. The maximum overshoot is

$$\theta_i \exp\left[\frac{-\zeta\pi}{\sqrt{1-\zeta^2}}\right] = 0.4443 \text{ (plus final value of 1).}$$

To find the third overshoot, put $n = 4$

$$\frac{P_{k+n}}{P_k} = e^{-n\pi\zeta/\sqrt{(1-\zeta^2)}} \quad \therefore P_5 = 0.01732 \text{ (plus 1)}$$

$$\text{Error} = e^{-\zeta\omega_n t} \quad \therefore e^{-2.5st} = 0.02 \text{ (2\%)}$$

$$t = \frac{\ln 0.02}{-2.5} = 1.5648 \text{ s.}$$

8. See Figure 11.8 on the next page.

The gain margin is infinite because the second order characteristic does not cross the $-180°$ axis. The phase margin $= 180° - 165.3° = 14.7°$.

9. (i) Put $d = 0$ and take x as the input

$$\frac{y}{x} = \frac{20}{1 - 20H} \quad \therefore y' = \frac{20x}{1 - 20H}$$

put $x = 0$ and take d as the input

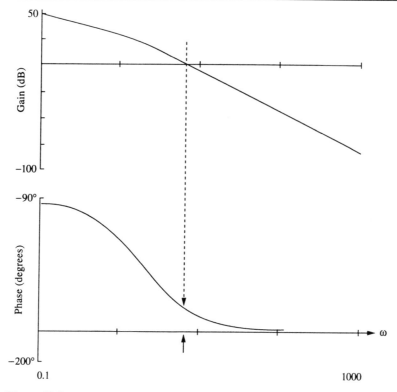

Figure 11.8

$$\frac{y}{d} = \frac{1}{1-20H} \quad \therefore y'' = \frac{d}{1-20H}$$

$$y = y' + y'' = \frac{20x+d}{1-20H}.$$

(ii) Put in the values

$$y = \frac{20(0.2)+0.02}{1-20(0.1)} = -4.02.$$

10. $\dfrac{V_o}{V_i}(j\omega) = \dfrac{1}{1+j\omega\tau} \quad (\tau = CR = 2 \text{ ms})$

when $\omega = 1/\tau$ then

$$V_o = V_i \cdot \frac{1}{1+j} = \frac{V_i \angle -45°}{\sqrt{2}} \text{ V}.$$

11. Take Laplace transforms and rearrange to give the transfer function

$$s^2\theta_{o(s)} + 4s\theta_{o(s)} + 100\theta_{o(s)} = 100\theta_{i(s)}$$

$$\frac{\theta_o}{\theta_i}(s) = \frac{100}{s^2 + 4s + 100} \quad \left[\text{compare to} \quad \frac{k\omega_n^2}{s^2 \zeta\omega_n s + \omega_n^2}\right]$$

$$\omega_n = 10 \text{ rad s}^{-1} \text{ and } \zeta = 0.2$$

$$\theta_{o(t)} = \theta_{i(t)}\left[1 - \frac{e^{-\zeta\omega_n t}}{\sqrt{1-\zeta^2}}.\sin\{\omega_d t + \phi\}\right]$$

$$\phi = \cos^{-1} 0.2 = 1.3694 \text{ rad}$$

$$\omega_d = \omega_n \sqrt{(1 - \zeta^2)} = 9.6824584 \text{ rad/s}$$

put values into the following equation

$$\theta_{o(t)} = \frac{\pi}{3}\left[1 - \frac{e^{-1}}{\sqrt{1-0.25}^2}.\sin\{6.21\}\right] = 1.076 \text{ rad.}$$

12. (a) See Figure 11.9.
    (b) See Figure 11.10.
    (c) See Figure 11.11.
    (d) See Figure 11.12.

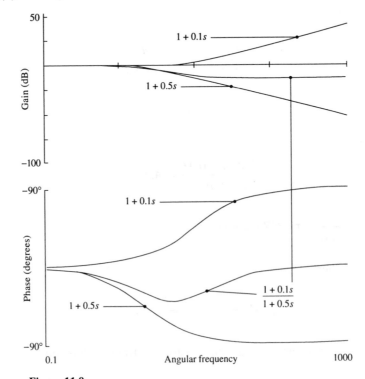

Figure 11.9

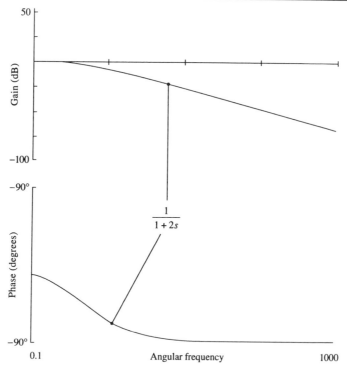

Figure 11.10

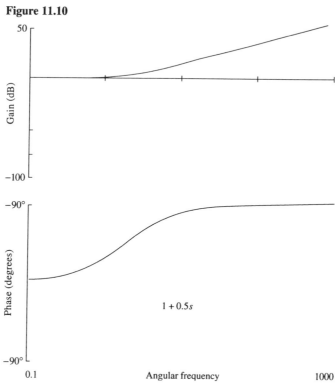

Figure 11.11

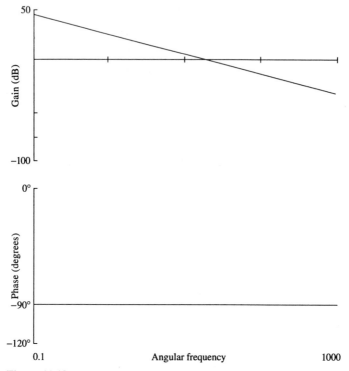

Figure 11.12

13. $\dfrac{V_o}{V_i}(s) = \dfrac{10+s6}{10+s6+s2} = \dfrac{10+s6}{10+s8}$

put $s = j2.5$

$$\dfrac{V_o}{V_i}(j\omega) = \dfrac{10+j15}{10+j20} = \dfrac{18.028\angle56.3°}{22.36\angle63.435°}$$

$$= 0.8063\angle-7.135°.$$

14. (a) (i) $\quad L(3\sin 5t) = \dfrac{3(5)}{s^2+25} = \dfrac{15}{s^2+25}.$

     (ii) $\quad L(1-e^{-t/\tau}) = \dfrac{1/\tau}{s(s+1/\tau)}.$

     (iii) $\quad L\dfrac{1}{3}e^{-2t} = \dfrac{1}{3}\left[\dfrac{1}{s+2}\right] = \dfrac{1}{3(s+2)}.$

(iv)
$$L(10\sin 6t + 20\cos 5t) = \frac{10(6)}{s^2 + 36} + \frac{20(s)}{s^2 + 25}$$

$$= \frac{20s^3 + 60s^2 + 720s + 1500}{(s^2 + 36)(s^2 + 25)}.$$

(b) (i)
$$L^{-1}\left[\frac{3+2s}{s(s+3)}\right] = \frac{A}{s} + \frac{B}{s+3}.$$

Using the cover-up rule $A = 1$ and $B = 1$. Hence

$$F_{(s)} = \frac{1}{s} + \frac{1}{s+3}$$

and $f(t) = 1 + e^{-3t}$.

(ii) Put in values and

$$F_{(s)} = \frac{4}{s(0.65+1)} = \frac{A}{s} + \frac{B}{0.65+1}.$$

using the cover-up rule $A = 4$ and $B = -2.4$

$$\therefore F_{(s)} = \frac{4}{s} - \frac{2.4}{0.65+1} = \frac{4}{s} - \frac{4}{s+1.67}$$

$$f(x) = 4 - 4e^{-1.67t} = 4(1 - e^{-1.67t}).$$

(iii)
$$F_{(s)} = \frac{2}{3s+2} = \frac{2/3}{s+2/3}$$

$$L^{-1} = \frac{2}{3}e^{-(2t/3)}.$$

(iv) There are several methods of solving this problem. The quickest is to use the Laplace transform pair

$$\frac{1}{s^2 + 2as + \omega_n^2} \Rightarrow \frac{1}{\omega_d}e^{-at}\sin\omega_{dt},$$

where $\omega_n = \sqrt{(\omega_d^2 + a^2)}$ or we can transpose the expression using the first shift theorem once we have completed the square on the denominator and then use the tables.
    We will try both methods

(i)
$$\frac{4}{s^2 + 6s + 100}\text{compare to }\frac{1}{s^2 + 2as + \omega_n^2}$$

we can deduce a scale factor of 4, also $a = 3$ and $\omega_n = 10$. Hence $\omega_d = \sqrt{(\omega_n^2 - a^2)} = \sqrt{91}$ rad/s. put these values into the inverse transform and

$$f(t) = \frac{4}{\sqrt{(91)}} e^{-3t} \sin \sqrt{(91)}t.$$

(ii) Complete the square on the denominator

$$s^2 + 6s + 100 \text{ (multiply by 4)}$$
$$4s^2 + 24s + 400 \rightarrow 4s^2 + 24s = -400$$

(add 36 to both sides)

$$4s^2 + 24s + 36 = -364$$
$$(2s + 6)^2 + 364 = 0$$
$$4(s + 3)^2 + 364 \text{ therefore } (s + 3)^2 + 91.$$

Hence the

$$F(s) = \frac{4}{(s+3)^2 + 91}$$

put $(s + 3) = s$ and multiply $f(t)$ by e^{-3t} (first shift theorem)

$$\frac{4}{s^2 + 91} \text{ gives } \frac{4}{\sqrt{(91)}} \sin \sqrt{(91)}t.$$

hence $f(t)$ is $4\sqrt{91}$, $e^{-3t} \sin\sqrt{(91)}t$ as above.

15.
$$\frac{\theta_o}{\theta_i}(s) = \frac{10}{2s+3} \text{ put } s = j6$$

16. For a first-order circuit, with time constant CRs, the current will decay from a maximum of V/R at an exponential rate, i.e.

$$i = \frac{V}{R} e^{-(t/\tau)} = \frac{30}{20k} e^{-(0.05/0.04)} = 0.43 \text{ mA}$$

17. $\theta_i = 5V$, $\theta_o = \pi/9$

$$i = \frac{\theta_o}{\theta_i} = \frac{\pi}{45} \text{ rad/V.}$$

18. (i) Open loop

$$= \frac{100}{(1+0.05s)(1+0.2s)}$$

put $s = j50$

$$\frac{100}{j^2 25 + j12.5 + 1} = \frac{100}{-24 + j12.5} = \frac{100}{27.06\angle152.5°} = 3.695\angle - 152.5°.$$

(ii) Closed loop (assume negative feedback)

$$= \frac{100}{0.01s^2 + 0.25s + 101} \left[\text{using } \frac{G}{1+G}\right]$$

put $s = j50$ as before

$$\frac{100}{25j^2 + 12.5j + 101} = \frac{100}{76 + j12.5} = \frac{100}{77.02\angle9.34°}$$

$$= 1.3\angle - 9.34°.$$

Index